Learning Centre

Park Road, Uxbridge Middlesex UB8 1NQ
Renewals: 01895 853326 Enquiries: 01895 853344

Please return this item to the Learning Centre on or before this last date stamped below:

2 1 APR 2006	- 7 JUN 2018	
0 5 OCT 2006		
2 5 MAY 2007		
0 8 JAN 2009		
2 4 JUN 2009		
2 2 OCT 2009		
0 7 OCT 2016		

CITB
Essential electrics

Published by
CITB-ConstructionSkills,
Bircham Newton, King's Lynn,
Norfolk PE31 6RH

First published 1994
Revised 1995
Revised 1996
Revised 2003
Revised 2005

© The Construction Industry Training Board 1994

ISBN 1 85751 026 7

Printed in the UK

CONTENTS

ACKNOWLEDGEMENTS

CITB-ConstructionSkills wishes to express its thanks to the following for their assistance in the production of this manual.

Consultant

Frank Saxton

Others

Mike Roper of Elm Training Services and Nottingham AM2 Test Centre

Martin Edwards of Training by ME (ME Technical Services)

Martin Hoare of Techtrain Associates

Polar Pumps Training

Note: This manual contains abbreviated extracts and paraphrases of the IEE regulations. It is emphasised that these interpretations of the regulations have been devised for the purpose of training and should not be regarded as authoritative in any other context. When necessary, the regulations should be referred to directly.

PREFACE

The matter of health and safety is never more important than when you are working on-site with other trades/personnel in the vicinity and when you are using a combination of pressurised gases, rotating equipment and electrical power.

It only needs a couple of seconds' carelessness for a major injury to occur with serious consequences. When you read this book, you may be very surprised to find out just how little electrical energy is required to cause severe injury or even death – not to mention material damage. The consequences of carelessness not only relate to health and safety but they also have severe legal implications.

The Health and Safety at Work Act 1974 gives strict rules regarding working practices. Some of the most important points are as follows:

1. The Act is designed to protect employers, employees and other personnel from injury or illness, whether they are in a factory, private house, public place or other area.

2. It should be remembered that both employers and employees may be deemed to be equally liable in the event of an accident or injury, and there are severe penalties at Magistrates' Court and also (to a far greater degree) at Crown Court.

The maximum penalties at Magistrates' Court are a fine of up to £5,000 or, in some cases, imprisonment.

The maximum penalty at Crown Court is an unlimited fine or a possible PRISON SENTENCE.

IMPORTANT

It is up to the employee as much as the employer to ensure that they take all sensible precautions to protect against accident/injury. This includes such items as safety glasses, ear defenders, safety hats, safety shoes, 110 V electrical equipment, etc.

The Health and Safety Executive (HSE) can bring about legal proceedings and has the power to shut down any operation that it considers is endangering the health and safety of any person.

It is essential that proper precautions are taken at all times, especially when you are working in public areas. It is no good putting up a notice warning of a danger if there are small children or blind people about. There must be physical barriers if there is a risk of any person straying into an unsafe area.

When using electrical equipment, it is essential that persons working on control panels or other live equipment are fully protected. Where necessary, supplies must be isolated safely and, where appropriate, a 'permit to work' system must be introduced.

When a certain competence is required to work on equipment, it is essential that the personnel working on that equipment are suitably trained. The phrase 'competent person' is explained in detail in the Health and Safety at Work Act 1974 documentation.

The Electricity at Work Regulations 1989 became effective on 1 April 1990. They were brought out to replace the Electricity (Factories Act) Special Regulations 1908 and 1944.

The Electricity at Work Regulations 1989 are statutory regulations under the Health and Safety at Work Act 1974, which are enforced by the Health and Safety Executive (HSE).

The Electricity (Factories Act) Special Regulations 1908 and 1944 were concerned with factories and gave protection to some six million people.

The Electricity at Work Regulations cover over 20 million people, imposing duties on persons, referred to as 'Duty Holders', with respect to:

- electrical systems
- electrical equipment
- electrical conductors
- work activities on or near electrical equipment.

They apply to all persons at work:

| There are no exceptions |

The legal duties imposed on employers, employees and the self-employed by the Electricity at Work Regulations are in addition to those already imposed by the Health and Safety at Work Act.

The regulations apply wherever the Health and Safety at Work Act applies and wherever electricity may be encountered. The Act contains some measures to prevent overlap and possible conflict with electrical regulations that apply to seagoing ships, aircraft, hovercraft and other vehicles, and such places as mines.

The purpose of the regulations is to prevent death or personal injury to any person from electrical causes in connection with their work. This may include:

- electric shock
- burns
- fires
- arcing
- explosions.

The regulations are divided into four parts:

- Introduction
- General
- Mines
- Miscellaneous.

This manual is concerned with Part Two – General only.

Part P to the Building Regulations for England and Wales. In May 2003 the government announced plans to include domestic electrical installation work under the legal framework of the Building Regulations. It was decided that Part P would be introduced when self-certifying schemes were in place to try to ensure the competency of electrical work undertaken. These schemes are now available and the requirements of Part P are effective from 1 January 2005.

BS 7671 (IEE WIRING REGULATIONS) DEFINITIONS

Exposed conductive part: A conductive part of electrical equipment that can be touched and is not a live part but may become live under fault conditions, e.g. steel conduit.

Extraneous conductive part: A conductive part liable to introduce a potential. This is generally an earth potential and it does not form part of the electrical installation, e.g. copper water pipe.

Direct contact: Contact of persons or livestock with live parts.

Indirect contact: Contact of persons or livestock with exposed conductive parts which have become live under fault conditions.

Live part: A conductor intended to be energised in normal use (including a neutral conductor).

System: An electrical system consists of a single source of electrical energy and an installation.

Distributor: A person who distributes electricity to consumers using electrical lines or equipment they own or operate.

1. ELECTRICAL CONCEPTS

Electricity

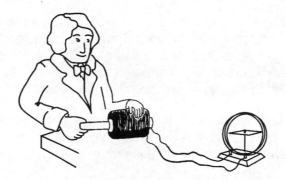

Michael Faraday discovered how to make electricity in 1831 when he plunged a bar magnet into a coil wire and thus generated a wave of electricity. He later found that, by rotating a copper plate between the poles of a magnet, power could be taken from the axis to the rim of the disk.

This system of holding the coil of wire stationary while varying the strength of the magnetic field is used in power stations. However, the bar magnet is replaced by a rotating electromagnet and the coils are arranged so that the windings are cut by the magnetic field as the magnet rotates.

Electrical charge

The basis of electrical energy is electric charge. There are two kinds of electric charge: positive and negative.

All materials consist of tiny particles called atoms. Materials which consist of the same kind of atom are called elements. The simplest atom is that of hydrogen. If we could see it, it would resemble the Earth with the Moon orbiting around it.

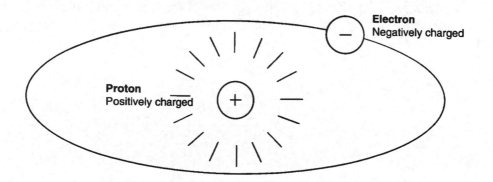

The two types of electric charge play a fundamental role in the atomic structure of matter.

Every atom consists of a nucleus of positively charged particles (called protons), around which negatively charged particles (called electrons) are spinning. Normally, the number of electrons in an atom equals the number of protons in the nucleus, and the atom is then said to be balanced.

If most of the atoms in the piece of material are electrically balanced, the material is said to be electrically neutral, or uncharged. Sometimes, material loses some of the electrons which belong to its atoms so that it contains more protons than electrons. It is then positively charged. Alternatively, an object may gain more electrons than are needed to balance all the protons in the atoms. The object is then negatively charged.

Electric charges share with magnetic poles the property of exerting forces upon one another. It must be remembered that like charges repel one another and unlike charges attract one another. This rule applies to the charges of the electron and proton.

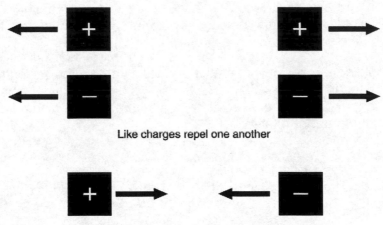

Like charges repel one another

Unlike charges attract one another

Since the nucleus is positive and the electron is negative, the electron is bound in orbit and is normally prevented from flying off because of the attraction between the two particles. The number of orbiting electrons in a given atom depends on the type of element.

The principle that like charges repel and unlike charges attract can be used to force electrons to move in the same general direction and thus to produce current flow.

This is achieved by an external negative charge (which is no more than a point with an abundance of electrons) placed at one end of a conductor and an external positive charge (a point with a deficit of electrons) placed at the other. This causes free electrons to flow towards the positive end, since electrons are negatively charged and unlike charges attract. Further, the external negative charge repels free electrons into the material so that there is ordered directional electron flow from negative to positive, as illustrated.

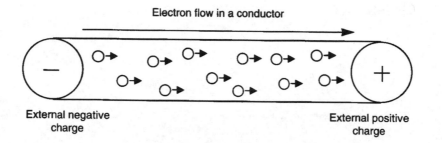

In conductive materials, the electrons not tightly linked to the atoms can be dislodged from their orbit by the force from an electric power supply. In this situation some of the negatively charged electrons are free to move and flow towards the positive terminal of the power supply.

Conductors and insulators

Materials in which the outer electrons are not tightly bound in the atoms and that can easily be dislodged to produce free electrons are called conductors (for example, copper and aluminium). Conversely, in some materials, the orbiting electrons are so tightly bound that they cannot easily be encouraged to break away from their orbits. These materials are called insulators. Insulators have virtually no free electrons available to form an electric current. Examples of insulators are plastics and ceramics.

Types of conductors and insulators

The most common types of conductors found in electrical installations are:

Copper: found in cable and flex
Brass: found in electrical accessories, such as terminal blocks
Nichrome: found in electric fire elements.

The most common type of insulator used in electrical installations is plastic, of which polyvinyl chloride (PVC), a thermoplastic, is the most widely used. Its wide use is due to its ability to be plasticised, which results in a range of flexible plastics, from rigid to pliable. The softer material is used as insulating covering for electric cables and wiring.

Free electrons

Consider a copper atom illustrated in the following diagram. It has 29 orbiting electrons arranged in four shells.

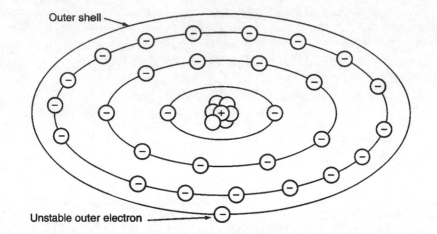

The outer electron, farthest from the nucleus, is only weakly attracted to the positively charged nucleus. This means that it can easily fly off or be dislodged. Electrons dislodged from their orbit can wander at random from atom to atom within the material and so are called free electrons. These free electrons can form the basis of an electric current.

The random movement of electrons

Under normal conditions, free electrons move randomly in a conductor; the effects of temperature cause this movement (see illustration).

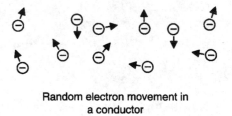

Random electron movement in
a conductor

This movement is usually equal in all directions so that no electrons are given up by the material and none are added. This is not electric current. However, if the free electrons can be encouraged to move in the same general direction within the conductor and can be made to enter and leave it, this flow does constitute an electric current.

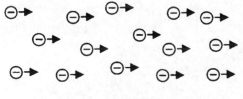

Directional movement – electric
current

Potential difference (electric pressure)

Electrical energy is stored in material when it is either positively or negatively charged. The energy is contained in the charge because, in order to produce it, electrons have to be forced together against the forces of repulsion which exist between them. The amount of charge stored in any given piece of material depends upon how many electrons it has lost or gained. Because electrons are very small, many millions have to be displaced to produce a measurable charge. The extent to which an object is charged is termed 'potential'.

Potential difference is measured in volts.

Voltage is the force behind electricity. It is often referred to as electric pressure and can be readily compared with the water pressure in a plumbing system. The pressure which drives the water is due to the difference of levels between the tank and the tap. The difference in the voltage levels between two points is called the potential difference (p.d.).

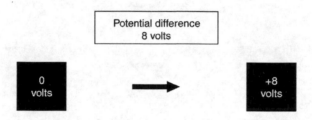

Charge passes across a potential difference only if the two objects are connected by a material which allows electricity to pass through it. Normally, the passage of charge takes the form of a movement of negatively charged electrons but, sometimes, there is a two-way movement of positively and negatively charged particles. When a connection is made between two charged objects, charge passes until the two objects reach the same potential, at which point all movement of charge stops.

Any equipment that works by electricity needs a power supply. Different equipment needs different types of supply. A torch may require two 1.5 V batteries, while an electric shower needs a 230 V AC supply.

The very high voltages that are used on power lines carried by pylons are measured in kilovolts (symbol kV), where one kilovolt is equal to 1,000 V.

Another term for voltage is EMF (electromotive force).

Electric current (amperes)

When charge moves from one place to another, an electric current is said to flow. Electric current is always regarded as flowing from the more positively charged object to the more negatively charged object.

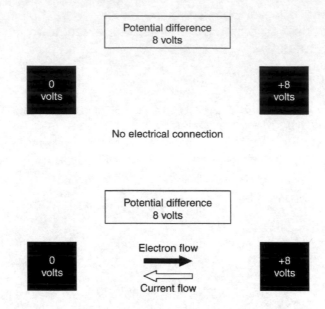

In a conventional circuit, an electric current (unlike water) will only flow if it can return to its source. The route it takes is known as a circuit. If you break a circuit by cutting a wire that forms that circuit, the current stops.

The unit of current is the ampere, and this is measured in amps (A). An immersion heater takes around 12 A, while the electronic components in a boiler may take only 1 milliamp (mA), which is one thousandth of an amp.

Note: Voltage appears across components and current flows through them.

Resistance

If electric current is like a flow of water, the path it flows along, which is the electrical circuit, can be likened to a heating system with obstacles in its path, such as radiators, that reduce the flow of water in the system.

In electrical circuits even the circuit conductors provide some degree of resistance to the flow of current. This is why voltage is always needed to push the current around the circuit to overcome the resistance.

Consider any piece of equipment – for example, a 110 V electric shaver. If it is accidentally connected to a 230 V supply system, too much current will flow through it and it may burn out. The solution would be to fit some kind of resistance in the circuit to limit the current. This is what happens when electric shavers are made to work on either 230/110 V systems.

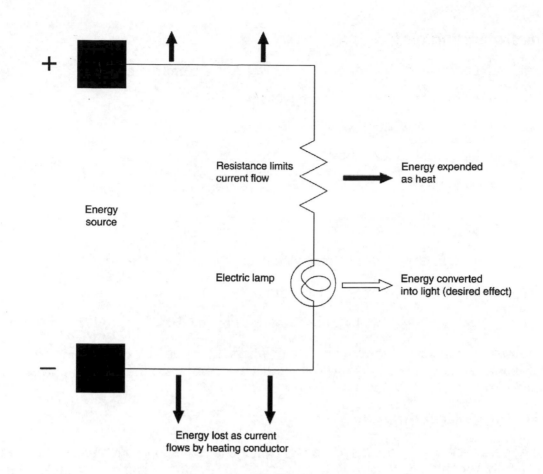

When an electric current meets resistance in a circuit, it generates heat by working its way through the conductor. This heating effect can be put to a practical use – for example, when resistance elements are used in electric fires and also in the filament of a lamp, where the heat produced is enough to make the filament white-hot.

Resistance can cause problems in a circuit, since overheating is a major cause of electrical breakdowns and can give rise to fire risks. If, when terminating conductors into electrical accessories or flexible cords into plug-tops, the terminal screw is not tightened sufficiently, the resulting high-resistance connection will get hot and will damage the accessory. Another factor to consider is the length of cable: the greater the length of run, the greater the resistance, and hence the greater the heat created. Care needs to be taken to ensure that any heat created in cables due to applied load and conductor resistance does not damage the insulation.

The resistance of a component or part of a circuit is measured in ohms, and the omega symbol (Ω) is used to represent it.

Typical values of resistance used in electrical installation and maintenance work are:

ohms 1 k = 1,000 Ω or 1 kilohm

1 M = 1,000,000 Ω or 1 megohm

Understanding electricity

To gain an understanding of electricity, think of it as a plumbing system:

- the height of the tank is voltage (volts)
- the water in the pipework is current (amps)
- the tap is resistance (ohms).

For a given voltage:

- the more you open the tap (less resistance) the more the water flows (more current)
- if you close the tap (more resistance) less water flows (less current).

From this simple analogy it can be seen that:

VOLTS push **AMPS** through **OHMS**

If the water from the tap flows over a water wheel, the speed of the wheel would represent electrical power (watts). The more water that flows, the faster the wheel would turn and therefore the more power.

Relationships (Ohm's law)

If the resistance of a circuit is high, a high voltage is required to push the current round the circuit. When the voltage falls and the resistance of the circuit remains the same, there is less current. From this it is evident that the voltage, current and resistance in a circuit are related to each other. This relationship is known as Ohm's law, after George Ohm, a German physicist.

Expressed mathematically:

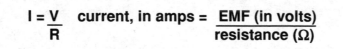

$$I = \frac{V}{R} \quad \text{current, in amps} = \frac{\text{EMF (in volts)}}{\text{resistance } (\Omega)}$$

alternatively, $V = I \times R$ or $R = \frac{V}{I}$

In terms of a simple circuit of a battery and flashlamp bulb, if the battery has an EMF of 12 V and the bulb a resistance of 4 ohms, a current of 3 A will flow when the switch is closed.

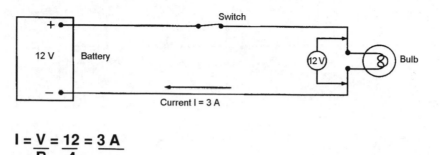

$$I = \frac{V}{R} = \frac{12}{4} = \underline{3\ A}$$

Series circuits

In the previous example we were concerned with a simple direct current circuit containing only a battery and lamp. Some circuits contain a number of items connected in series – that is to say, the same current passes through each item, in sequence.

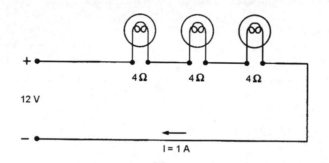

In the above circuit the same current passes through each lamp in turn. If each lamp has a resistance of 4 ohms, the total resistance of the circuit would be 12 ohms. With an EMF of 12 V, it is obvious (applying Ohm's law) that a current of 1 A would flow, and that this 1 A current would be common to all three lamps.

The total EMF of 12 V would be distributed across the total load of all three lamps. If all the lamps were the same (with the same current and power rating factors), each lamp would, in effect, have 4 V across its terminals.

This can be proved by applying Ohm's law, expressed as follows:

Voltage across lamp $\quad V = I \times R$
$\qquad\qquad\qquad\qquad 4\ V = 1\ A \times 4\ \Omega$

If the lamps did not all have the same resistance, different voltages would be developed across the terminals of each lamp. For example:

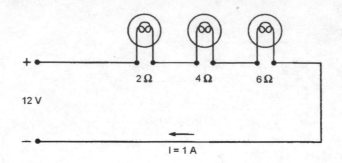

The total resistance of the circuit is still 12 ohms, and a common current of 1 A will flow through each of the lamps in turn. However (from Ohm's law) we can see that the voltage across each lamp would be different. For example:

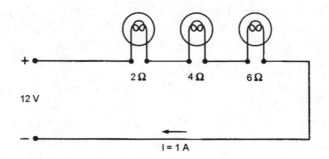

Potential difference (p.d.)

The voltage across each of the lamps is the potential difference (p.d.) required to sustain the common current (in this case of 1 A).

If a similar circuit is devised, but the lamps are replaced with resistors in series, different voltages can be taken at different points. For example:

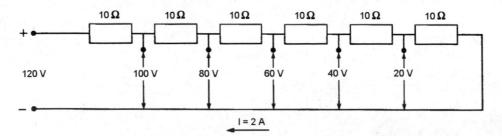

The total resistance of this circuit is 6 x 10 ohms = 60 ohms. If the input voltage is 120 V, a current of 2 A will flow through the circuit. The total voltage of 120 V will be 'divided' across all six resistors. Since they all have the same value (10 ohms), the 'voltage drop' or 'potential difference' across each would be 20 V and the voltage, measured with respect to the common negative, would vary in 20 V steps.

This type of circuit is called a 'potential divider' network and is commonly used where different or varying voltages are required.

Variable resistance

Where a varying voltage is required, this can be achieved by placing a variable resistor in the circuit. For example:

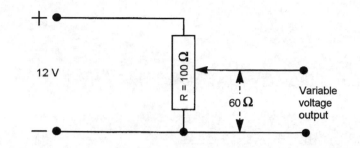

In this type of circuit, the output voltage (with no load connected) is 'tapped off' according to the position of the wiper arm. With an input current of 0.12 A flowing, 7.2 V will be tapped across the upper portion of the resistor, leaving 4.8 V across the output terminals. This type of circuit could be used to provide a reference voltage for, say, a temperature control circuit.

Parallel circuits

Elements in an electrical circuit may also be connected 'in parallel' – that is, the same EMF is applied to each element but the current flowing in each will vary, depending on its resistance. The greater the resistance, the less the current in each element.

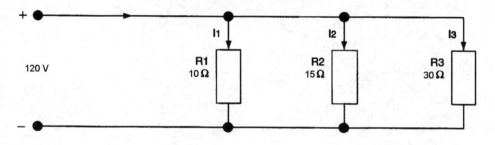

The effective resistance of such a circuit is given by the formula:

$$\frac{1}{R_t} = \frac{1}{R_1} + \frac{1}{R_2} + \frac{1}{R_3} \quad \text{etc.}$$

$$\frac{1}{R_t} = \frac{1}{10} + \frac{1}{15} + \frac{1}{30}$$

$$\frac{1}{R_t} = \frac{6 + 4 + 2}{60}$$

$$\frac{1}{R_t} = \frac{1}{5} \qquad \therefore \ R_t = \frac{5}{1} = 5\,\Omega$$

Note: R_t = Total resistance.

The current flowing through each element would be:

$$1_1 = \frac{V}{R_1} = \frac{120}{10} = \underline{12\,A} \qquad 1_2 = \frac{V}{R_2} = \frac{120}{15} = \underline{8\,A} \qquad 1_3 = \frac{V}{R_3} = \frac{120}{30} = \underline{4\,A}$$

If these were added together the total current (It) flowing through the circuit would be:

It $= 1_1 + 1_2 + 1_3$

It $= 12\,A + 8\,A + 4\,A$

It $= \underline{24\,A}$

which confirms the solution above.

Note: It = Total current.

In practice, many devices are connected in parallel and it is important to know the extent of the current flowing in each.

Combined series and parallel circuits

Many circuits include combinations of series and parallel circuits. An example is shown below:

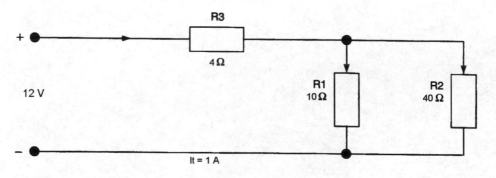

The effective resistance and actual current flow can be determined by employing the principles and formulae discussed previously.

The resistance of the parallel element is determined by the formula:

$$\frac{1}{R_t} = \frac{1}{R_1} + \frac{1}{R_2}$$

$$= \frac{1}{10} + \frac{1}{40}$$

$$= \frac{4 + 1}{40}$$

$$= \frac{5}{40}$$

$$\therefore \underline{R_t = \frac{40}{5}} = \underline{8\ \Omega} \qquad\qquad \therefore R_t = \underline{8\ \Omega}$$

The parallel element is effectively in series with R_3 so that the effective resistance of the entire circuit is:

$4 + 8 = 12\ \Omega$

From Ohm's law, we can determine that the current flowing in the circuit would be 1 ampere. The voltage drop across the series resistance is 4 V, with 8 V across the parallel element.

We can now determine the current flowing in each part of the parallel element, again by the application of Ohm's law:

$$I_1 = \frac{8\ V}{10\ \Omega} \qquad\qquad I_2 = \frac{8\ V}{40\ \Omega}$$

$$I_1 = \underline{0.8\ A} \qquad\qquad I_2 = \underline{0.2\ A}$$

The total current passing through the parallel element is $0.8 + 0.2 = 1$ A, confirming the result previously determined.

Electric power

In order to do its work, electricity generates power. Power is the rate at which electrical energy is converted into other kinds of energy, such as heat, light or movement (in the case of electric motors).

The unit of power is the watt, and typical values of power used in electrical circuits are:

Kilowatt = 1,000 watts or 1 kW
Megawatt = 1,000,000 watts or 1 MW

Many electric motors are rated in horsepower – 1 h.p. = 746 watts. Electrical power can be calculated by multiplying the volts by the amps:

Watts (P) = volts (V) x amps (I)

It must be remembered that the formula applies accurately only to DC supplies, but it can be used for rough calculations for AC circuits:

Watts (P) = voltage x current

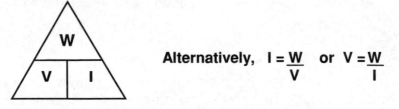

Alternatively, $I = \dfrac{W}{V}$ or $V = \dfrac{W}{I}$

Example: What is the actual current taken by a 3 kW immersion heater, if the supply system is 230 V?

$$I = \frac{W}{V} = \qquad I = \frac{3,000}{230} \qquad I = 13\ A$$

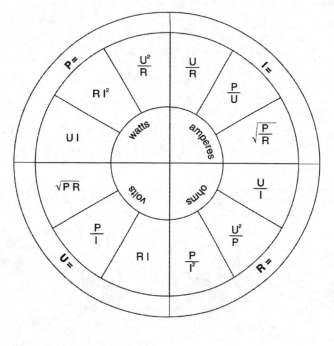

Symbols
U = Voltage in volts
I = Current in amperes
R = Resistance in ohms
P = Power in watts

Types of power supply

There are two types of power supply: alternating current (known as AC) and direct current (known as DC). Alternating current is the type of electricity supplied to domestic, commercial and industrial premises by the local electricity distributor. Direct current is the type of electricity you get from a battery or by using certain components to form a special type of circuit. It is possible to change AC to DC and vice versa.

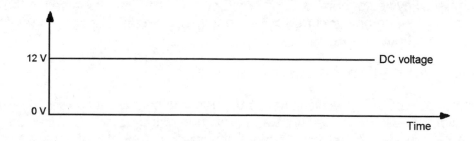

AC theory

All the previous illustrations have related to current which flows in one direction only. The current flowing in a circuit can, however, follow a different pattern (i.e. alternating or AC current), which is shown in the diagram on the next page.

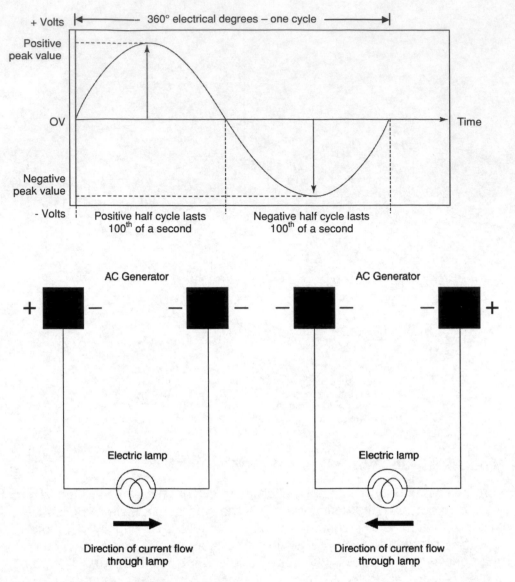

The number of cycles of current flow which occurs in one second is called the frequency of the alternating current. A frequency of 50 Hz is standard for electricity supplies throughout the UK.

Advantages of alternating current

Generators are the only practical means of supplying continuous power. The relative simplicity of the AC generator (or alternator) is one of the advantages of using an AC electrical system. AC, however, has other advantages over DC, which tend to reduce the costs of power production. Among the most important are:

1. some types of electrical apparatus designed to operate from AC supplies are much simpler and easier to maintain than similar apparatus designed to work from DC supplies (in particular, alternating current motors)

2. the voltage of an AC supply can be changed very easily. For example, a low voltage supply can be obtained from a higher voltage supply by means of a transformer (but there is no simple method of changing the voltage of a DC supply)

3. electricity can be distributed more easily and efficiently with an AC supply than with a DC supply. Transformers make it possible to distribute electricity at high voltages. The higher the voltage of transmission, the greater the efficiency due to the fact that they carry less current for a given amount of power and, consequently, the losses or volt drop is reduced.

RMS current and voltage

The root mean square value is the value most commonly used when expressing mains voltage as a value of 230 V.

In order to explain what this means we can use, as an example, a 100 W bulb which glows at a fixed brightness when connected to the mains supply.

The direction of current flowing through the filament plays no part in determining the brightness. This is determined by the power generated within the filament. The current flowing through the filament and, hence, the power level are continually changing as the supply voltage changes. When the mains waveform reaches a peak value, maximum current and hence maximum power are developed in the load. Similarly, as the sine wave voltage reaches zero, no current and therefore no power is generated in the bulb.

If this is the case, why do we not see the brightness vary? The answer is that it would if the frequency of the mains was much lower than 50 Hz. The power and heat are effectively averaged out during each cycle.

To produce the same average power in the load, the peak value must be higher than the RMS value. The relationship between the peak and RMS values has long been established and is illustrated below.

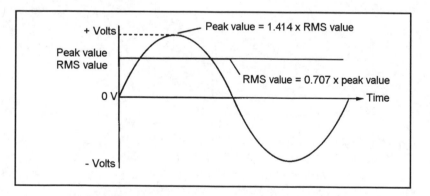

This relationship can be proved by a mathematical process, but this is beyond the scope of this publication.

Electromagnetic induction

Electrical energy is produced in a conductor by the magneto-electric effect created when a conductor is moved in a magnetic field. The creation of a voltage in a conductor by this means is termed 'induction'. We say that a current is induced in a conductor, provided that the conductor forms part of an overall circuit.

It follows that a similar result can be achieved if the conductor is stationary and the magnetic field is moved (i.e. a current is induced in the conductor as a result of the magnetic lines of flux cutting the conductor).

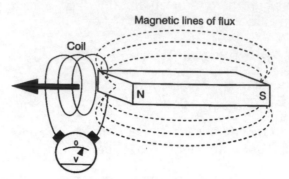

Mutual induction

In the illustration, a permanent magnet provides the magnetic lines of flux to cut the conductor. The permanent magnet could be replaced by an electromagnet. In this case, a current is passed through a coil to produce a magnetic field and the necessary flux to induce a voltage across the conductor. When a current in one coil induces a current in another adjacent coil, this is termed 'mutual induction'.

Transformers

Transformers are inductive devices which use the properties of mutual induction. This means that they rely on a continually changing primary flux to permit an induced voltage to be developed across the secondary. Therefore they can only be used to transform AC voltages; they cannot work at all in pure DC circuits where the primary current is kept constant.

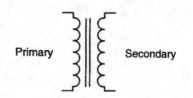

Transformers are used as voltage conversion devices – they change the level of one AC voltage to another, either upwards or downwards. Alternatively, they can be used as isolation devices, allowing two circuits to be coupled without there being a direct electrical connection, as in the case of bathroom shaver units.

If the number of turns on the primary winding equals the number of turns on the secondary, then the induced secondary voltage will equal the applied primary voltage. The transformer is then said to have a 1:1 turns ratio. In practice, losses within the transformer mean that the turns ratio only gives an approximate guide to the primary–secondary voltage relationship.

A turns ratio of 10:1 would produce a voltage across the secondary coil of one tenth of the input voltage (i.e. a 230 V input would produce a 23 V output). Transformers can also be used to 'step up' a voltage. A turns ratio of 1:4 would produce almost 1,000 V output from normal mains input.

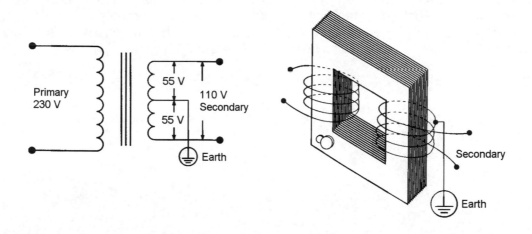

Secondary windings often have a number of graduated 'taps' which provide a variety of outputs. Some transformers include a centre tap in the secondary coil, which is usually earthed. This effectively halves the maximum output voltage relative to earth. In most cases the soft iron core and outer casing are bonded to earth. This type of transformer is used to supply 110 V portable tools on construction sites.

The rating of a transformer

With electrical equipment we must always consider the maximum current that can be carried without exceeding the rating. Transformers are limited in the amount of current they can supply from their secondary windings. If too much current is drawn, the windings get hot, which could melt the winding's insulation and thus burn it out.

Transformers are power-rated in volt-amps (VA). Watts cannot be used to represent power dissipation in a transformer because the voltage and current are not in phase with each other as they are in a pure resistance. By using a volt-amps figure (which is essentially the same, arithmetically speaking), it is relatively simple to calculate the current that can be drawn from a transformer.

When we say a transformer has a particular VA rating we are saying that this is the maximum power that can be drawn from the secondary/secondaries. Consider a mains step-down transformer that has a VA rating of 1 kVA and a single secondary winding of 110 V. What maximum current can safely be drawn from the secondary?

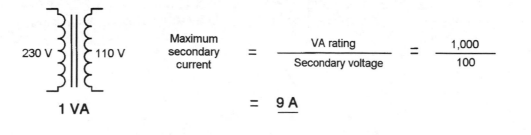

$$\text{Maximum secondary current} = \frac{\text{VA rating}}{\text{Secondary voltage}} = \frac{1{,}000}{100}$$

$$= \underline{9\ A}$$

Capacitance

A capacitor (sometimes referred to as a condenser), in its simplest sense, is a device for the temporary storage of electrical energy. It comprises two parallel metal plates, insulated from each other. If a DC voltage is connected across them, one of the plates becomes rich in electrons; the other plate becomes correspondingly poor. In acquiring this charge a current flows, but only for an instant. No sustained direct current can flow between the plates, since they are insulated, one from the other.

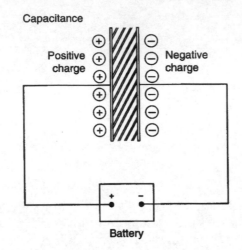

If the DC source is removed, the capacitor will retain its charge until it is discharged through an external circuit.

If an alternating current is fed to a capacitor it will commence to charge on one half-cycle but, as the voltage falls from its peak, it will attempt to discharge and to charge up again (in the opposite direction) on the next half-cycle, and so on. As a result, a capacitor appears to pass current when connected to an AC source, but prevents the passage of DC current.

The larger the area of the plates in a capacitor, the greater the capacitance. In practice, a capacitor is made from two thin sheets of metal foil, insulated by waxed paper, mica or similar material known as a dielectric.

The unit of capacitance is the farad. In practice this is far too big, and the micro-farad (one millionth of a farad) is the unit in common use. This is sometimes written as µF.

Rectifiers

It is sometimes necessary to convert an alternating current into a direct current. This is done by a process known as rectification. A rectifier is rather like a valve which permits current to pass freely in one direction but which prevents current passing in the opposite direction.

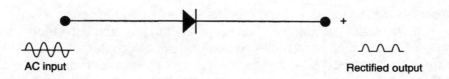

AC input   Rectified output

As we have seen, an alternating current is one which passes through a cycle from a positive peak through zero to a peak in the opposite (negative) direction. A simple rectifier would prevent current passing on the negative cycle but would permit the positive element to pass through. The resultant output would be a series of peak voltages.

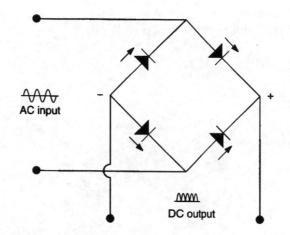

AC input

DC output

The rectifier may be a single, solid-state diode or a more complex, full-wave or bridge rectifier which is so arranged as to rectify both the positive and negative half-cycles of an alternating current, producing a DC output at twice the original frequency, which is easier to smooth (and twice the average voltage).

Rectifier devices such as these are commonly found in control equipment.

Filter circuits

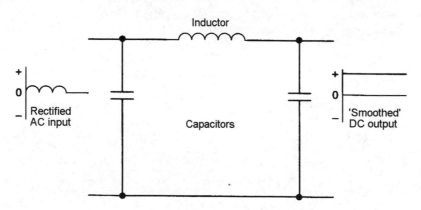

Inductor

Rectified AC input

Capacitors

'Smoothed' DC output

The output from a full-wave rectifier produces pulses of unidirectional current rather than a smooth DC output such as that produced by a battery. If smooth DC output is required, there are various types of filter circuit that can be added, from a simple capacitor to the more common circuit shown above.

The capacitors and inductors are both devices which will store energy when it is available from the supply and that will release it when the supply reduces, thus producing a smoothing effect. The two types of device complement each other because the capacitor works best at smaller values of load current and the inductor works best when large values of current are being supplied. This circuit will therefore provide a smooth output across a wide range of load currents.

Induction motor

The simplest of all electric motors is the squirrel-cage type of induction motor used with a three-phase supply. The stator of the squirrel-cage motor consists of three fixed coils. The rotor consists of a core, in which are imbedded a series of heavy conductors arranged in a circle around the shaft and parallel to it.

Three-phase induction motor

The stator

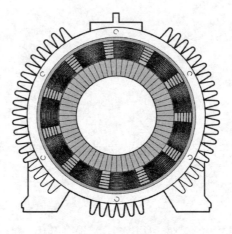

The stator's design is dictated by the number of field coils used per phase. In a three-phase motor there is a minimum of two coils per phase; the maximum is determined by the speed of the motor and the available space.

The rotor

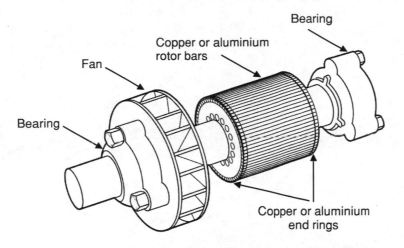

Cage rotor complete with fan and bearings

The rotor consists of a laminated core through which is fitted a cage of copper conductors and, because of its similarity to a cage, this construction is known as a 'squirrel-cage' rotor.

The bars of the cage are shorted together, forming a low-resistance circuit that consists of a number of single-turn coils.

The rotor sits in the space between the field coils, and any magnetic flux in the field windings envelopes the rotor.

Three-phase windings rotating fields

Introduction

Most electrical power generated is AC and, therefore, many motors are designed to operate on AC. The power supply could be single-phase or three-phase, but the majority of larger AC motors operate on a three-phase supply.

Principle of operation

The diagram on the next page illustrates the stator of a three-phase motor to which a three-phase supply is connected. The windings in the diagram are in a star formation and two windings of each phase are wound in the same direction.

Each pair of windings will produce a magnetic field, the strength of which will depend upon the current in that particular phase at any instant of time. When the current is zero the magnetic field will be zero, and maximum current will produce the maximum magnetic field.

As the currents in the three-phases are 120° out of phase (see the relationship in the diagram on the next page), the magnetic fields produced will also be 120° out of phase.

Rotating field of three-phase induction motor

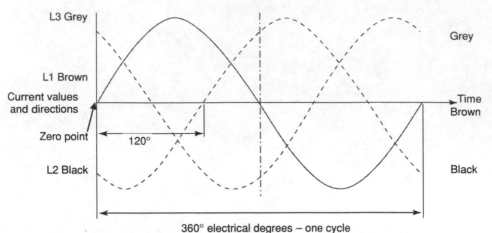

Rotating field of three-phase induction motor

The rotor of a three-phase induction motor turns due to the interaction of magnetic fields. As the rotating field of the stator cuts the rotor bars of the rotor, an EMF is induced into the conductors which form a closed circuit. A magnetic field is set up by the current flowing in the rotor conductors which interacts with the rotating magnetic field, causing the rotor to turn in the direction of the magnetic field.

A three-phase winding consists of three sets of coils connected in either 'star' or 'delta' formations evenly distributed around the stator core. Each winding could have two or more poles per phase depending upon the speed required. The magnetic field set up by a two-pole winding will complete one revolution in one cycle, while a four-pole winding will complete one revolution in two cycles of the supply. As the number of poles per winding is increased, so the speed of the rotating magnetic field within the machine decreases.

Motor speed

The synchronous speed of AC motors is dependent upon the frequency of the supply and the number of pairs of poles on the stator. The synchronous speed for any AC motor can be calculated from the formula:

$$N = \frac{f \times 60}{p} \text{ rpm}$$

When   N = speed in rpm

f = frequency in cycles/second (hertz)

p = number of pairs of poles

Therefore, a motor wound with two poles per phase and connected to a supply of 50 Hz would have a synchronous speed of:

$$N = \frac{50 \times 60}{1} = 3{,}000 \text{ rpm}$$

A four-pole motor would have a synchronous speed of 1,500 rpm. In practice, these synchronous and actual speeds are not achieved. The difference between synchronous and actual speed is called slip. Slip is due to magnetic losses in the gap between the stator and the rotor (the air gap) and it is proportional to the load on the motor.

Loss in speed is generally about 5% under full load conditions.

Motor speeds for three-phase machines

Number of poles		2	4	6	8	10	16
Sync speed	rpm	3,000	1,500	1,000	750	600	375
Approx FL speed	rpm	2,900	1,440	960	720	500	360

Terminal block markings

The leads from the windings will be either marked or coloured.

These ends may be coloured one colour for starting ends and a different colour for finishing ends, but BS 4999 now requires the markings U1, V1, W1, U2, V2 and W2 to be used:

U1	RED	V1	YELLOW	W1	BLUE
U2	BLACK	V2	BROWN	W2	WHITE

In older-type motors all the six leads may be the same colour (black) or the leads may be marked A, B, C or A1, B1, C1 and B2, C2, A2, etc., or sometimes U, V, W, X, Y, Z is used.

Reversal of rotation

The direction the magnetic field rotates is dependent upon the sequence in which the phases are connected to the windings. Rotation of the field can be reversed by reversing the connection of any two incoming phases.

Single-phase AC motors

A single-phase AC supply produces a pulsating magnetic field, not the rotating magnetic field produced by a three-phase supply. All AC motors require a rotating field to start.

With single-phase supplies, however, the benefit of 'phase shift' is not available, and the condition due to the field or stator winding is a magnetic flux which rises to maximum, falls to zero and then rises to a maximum in the opposite direction. This action produces a 'pulsating' magnetic field which rises and falls with the supply frequency but does not rotate around the stator.

The pulsating field of the stator cuts the rotor bars and induces magnetic flux into the rotor, but equal and opposite forces produced in the rotor prevent it from rotating since the net torque is zero.

One method of producing the necessary phase shift in the stator field is to introduce a second field winding known as a start winding. This is displaced on the stator by 90° to the main field winding, which is now known as the running winding. Both the stator and the rotor of AC motors are laminated.

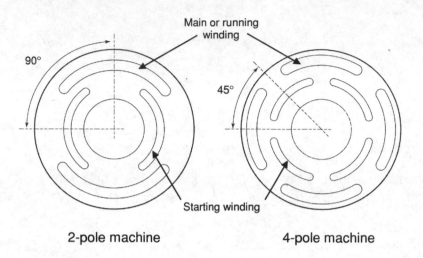

2-pole machine   4-pole machine

The split-phase motor

This is perhaps the most common of all induction motors as it is simple and cheap to produce while being suitable for a wide range of applications.

The running winding, which is designed to be highly reactive, is formed of heavier wire than the starting winding and has less resistance than the starting winding, which is made up of very fine wire. The currents in the two windings will therefore be out of phase with each other and reach maximum values at different times during each cycle. This being so, the magnetic flux due to each winding will rise and fall with the current and will also reach maximum and minimum levels at different instants of each cycle.

Because the windings are displaced around the stator, a movement of flux takes place between the start and running windings which cuts the rotor bars and sets the machine in motion.

The starting winding is only required for about a second at the most, and is automatically cut out by a centrifugal switch when the motor reaches about 75% full speed.

The moving arms for the centrifugal switch are fitted on the rotor shaft and operate contacts on the stator. Great care must be taken to ensure the correct alignment of this switch when reassembling the motor. Should the contacts fail to open, the start winding will burn out very quickly as the fine wire winding is incapable of carrying load.

To change the direction of rotation, reverse the connection of one winding only, i.e. start or run. (Standard practice is to reverse the start winding.)

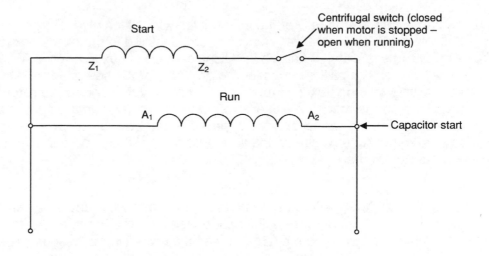

Capacitor start and capacitor start–run motors

In these type of motors, the start winding is given a leading power factor by the use of capacitors, which create flux movement. These capacitors are connected as shown below.

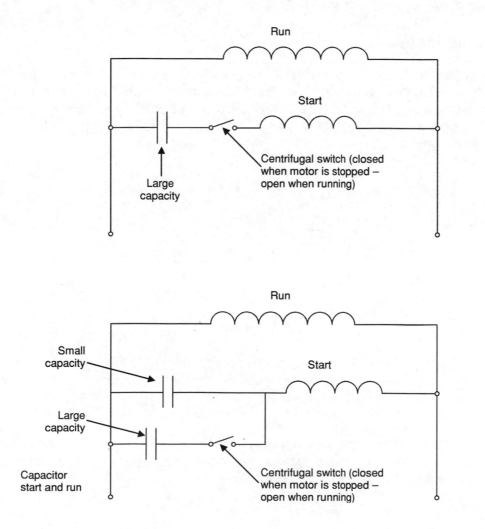

The capacitors are usually of the electrolytic type, which give a large capacitance for a relatively small physical size. They are normally fixed to the frame of the motor but can be mounted separately nearby. The displacement of flux in the two windings is much greater at starting for the capacitor motors than for the so-called 'split-phase' types. The starting winding flux is almost at a maximum value before the rotor begins to move, and so the machine has a better starting torque than the split-phase type. The capacitance of the capacitor fitted depends upon the starting torque the motor is required to produce but, once the motor is running, the same capacitance is neither required nor desired.

Manufacturers therefore offer two types of capacitor motor: the capacitor start machine, where a centrifugal switch disconnects both the capacitor and start winding on gaining speed, in the same manner as the split-phase motor; and the capacitor start–capacitor run machine, in which both a large and a small capacitor is used in the start winding for starting purposes.

The large capacitor is switched out by the action of the centrifugal switch, leaving the smaller capacitor in series with the start winding which, in this case, is capable of carrying load. The overall effect is a better running torque and an improved power factor.

The direction of rotation is changed as before by reversing the connections of either the start or running winding, but not both.

Shaded-pole motors

The shaded-pole motor is a simple, robust, single-phase motor that is suitable for very small machines with a rating of less than about 50 watts. Illustrated below is a shaded-pole motor. It has a cage rotor and the moving field is produced by enclosing one side of each stator pole in a solid copper or brass ring, called a shading ring, which displaces the magnetic field and creates an artificial phase shift.

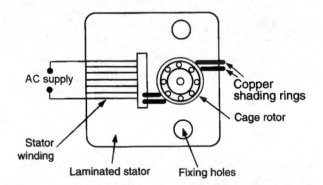

Reversal of rotation is theoretically possible by moving the shading rings to the opposite side of the stator pole face. In practice this is often not a simple process and, since the motors are symmetrical, it is sometimes easier to reverse the rotor by removing and fixing bolts and reversing the whole motor.

There are more motors that operate from single-phase supplies than all the other types of motor added together. Most are used in very small motors in domestic and business machines, where single-phase supplies are common.

Motor maintenance

All rotating machines are subject to wear, simply because they rotate. Motor fans which provide cooling also pull dust particles from the surrounding air into the motor enclosure.

Bearings dry out, drive belts stretch and lubricating oils and greases require replacement at regular intervals. Industrial electric motors are often operated in a hot, dirty, dusty or corrosive environment for many years. If they are to give good and reliable service they must be suitable for the task and the conditions in which they operate. Maintenance is required at regular intervals.

The solid construction of the cage rotor used in many AC machines makes them almost indestructible and, since there are no external connections to the rotor, the need for slip rings and brushes is eliminated. These characteristics give cage rotor AC machines maximum reliability with the minimum of maintenance and make the induction motor the most widely used in industry. Often the only maintenance required with an AC machine is lubrication, in accordance with the manufacturer's recommendations.

Generator electricity with a rotating magnetic field

A typical generator consists of copper conductors wound on to an armature which is rotated (for example, by an internal combustion engine, or a steam or water-driven turbine) within a magnetic field. However, it is easier to explain the principle of generation by describing the operation using a fixed stator winding and a rotating magnetic field, as shown below:

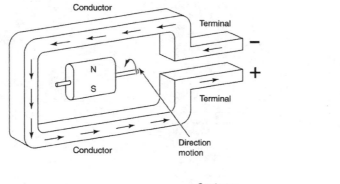

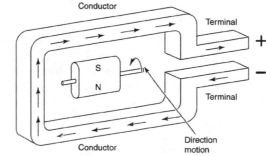

When the magnetic rotor is driven around, the magnetic field sweeps past all the conductors in the stationary part of the machine (stator) in turn. The direction of voltage and current flow reverses with each revolution.

The following illustrations
and diagrams show how
sine waves are generated.

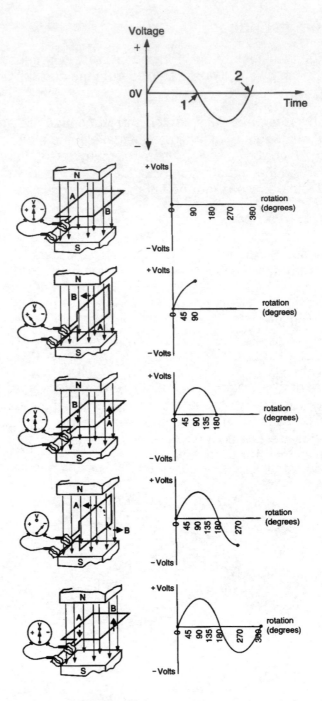

Alternating-current generators (alternators)

As stated above, a simple generator without a commutator will produce an electric current that alternates in direction as the armature revolves. Such an alternating current is advantageous for electric power transmission, and, hence, most large electric generators are of the AC type.

Low-speed AC generators are built with as many as 100 poles, both to improve their efficiency and to attain more easily the frequency desired. Alternators driven by high-speed turbines, however, are often two-pole machines. The frequency of the current delivered by an AC generator is equal to half the product of the number of poles and the number of revolutions per second of the armature.

34

It is often preferable to generate as high a voltage as possible. Rotating armatures are not practical in such applications because of the possibility of sparking between brushes and slip rings, and because of the danger of mechanical failures that might cause short circuits.

Alternators are therefore constructed with a stationary armature, within which revolves a rotor composed of a number of field magnets. The principle of operation is exactly the same as that of the AC generator described earlier, except that the magnetic field (rather than the conductors of the armature) is in motion.

The current generated by the alternators described above rises to a peak, sinks to zero, drops to a negative peak and rises again to zero a number of times each second, depending on the frequency for which the machine is designed. Such current is known as single-phase alternating current. If, however, the armature is composed of two windings, mounted at right angles to each other and provided with separate external connections, two current waves will be produced, each of which will be at its maximum when the other is at zero.

Such current is called two-phase alternating current. If three armature windings are set at 120° to each other, current will be produced in the form of a triple wave, known as three-phase alternating current. A larger number of phases may be obtained by increasing the number of windings in the armature, but in modern electrical-engineering practice three-phase alternating current is most commonly used, with the three-phase alternator the dynamoelectric machine typically employed for the generation of electric power. Voltages as high as 23,200 V are common in alternators.

2. SAFE ISOLATION

Introduction

Electricity, when safely controlled, is a very efficient and convenient way of distributing and using energy. If it is inadequately controlled, it can be lethal.

Electric shock

An electric shock effects the nervous system, causing the muscles to contract and sometimes concussion.

An electric shock can be received either by direct or indirect contact with electricity.

Direct contact – contact of persons or livestock with parts or conductors that are intended to be live in normal use.

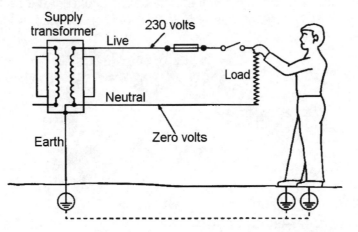

Indirect contact – contact of persons or livestock with exposed conductive parts that have become live under fault conditions.

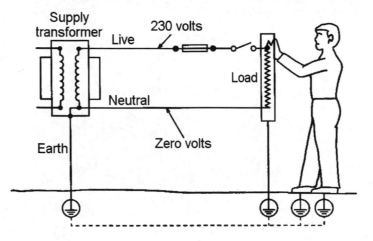

Body resistance

This varies depending on the conditions of the person's body, their age, situation and the weather.

Perception level

1 mA is the point at which an individual is aware of an electric current.

Let-go level

9 mA is the point at which an individual still has control over the effects of a shock on the muscles in their body and is thus able to release a gripped conductor.

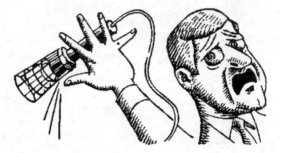

Freezing and muscular contraction

With further increases in levels of current (around 20 mA), individuals cannot release themselves. Extreme pain is felt, which may cause the subject to lose consciousness, or the body muscles may contract, affecting the lungs, and the subject may thus die from asphyxia.

Death

With currents of around 80 mA, death is likely to occur as a result of ventricular fibrillation and severe internal and external burns.

Ventricular fibrillation

A severe electric shock can cause the muscles of the heart to contract separately at different times instead of in unison. This condition (ventricular fibrillation) is a killer.

Electrocardiogram showing
normal heartbeat

Electrocardiogram showing
ventricular fibrillation

Removing persons

Great care needs to be taken when removing a person who has come in contact with live conductors. The following points should be considered:

- the rescuer must not put themselves in danger

- all necessary procedures must be undertaken as quickly as possible

- all necessary procedures must be carried out in a way that prevents further injury.

First, disconnect the electricity supply, wherever possible. Pull the victim away from the live conductors using dry clothing (such as overalls) that has been wrapped around them so that they can be removed effectively and quickly.

Treatment

Summon assistance and call for an ambulance. In the case of slight shock, reassure the patient and make them comfortable. Report the accident to the appropriate personnel.

If burns have been sustained, cool the areas with cold water or any other non-flammable fluid at hand. Remove anything of a constrictive nature if possible, such as rings, belts and boots. If the burns are serious, cool the areas and send the patient to hospital without delay.

For severe cases of shock where the patient is unconscious and not breathing, clear the airway and administer mouth-to-mouth resuscitation. Remember, there is no time to waste because a lack of oxygen to the brain can cause damage within four minutes.

When the patient starts to breathe again, place them in the recovery position and send them to hospital without delay.

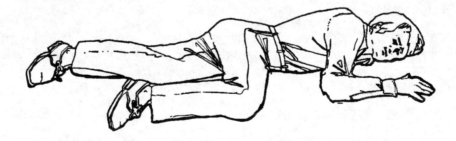

If the heart has stopped, then cardiac compression should be given.

Electrical isolation

Before beginning work on any electrical circuit, you should make sure that it is completely and safely isolated from the supply.

Electrically-powered machines are usually fitted with an isolator for disconnecting the supply under 'no load' conditions. It will be necessary to isolate the supply by either locking-off or removing fuses. The Electricity at Work Regulations 1989 definitely prefer locking-off as a means of isolation, but will accept the removal of fuses. The best method is to lock-off AND remove fuses.

Fuses, switch-fuses and isolators (i.e. equipment for isolation) should be clearly marked to indicate the circuit they isolate or protect.

Any isolating device, when operated, should be capable of being locked in the OFF or OPEN position, or carry a label stating otherwise. If the isolator consists of fuses, these should be removed to a safe place where they cannot be replaced without the knowledge of the responsible person concerned. For example, they can be kept in the pocket if the job is of short duration, or in a locked cupboard provided for the purpose in the charge of the works or site supervisor.

Fuses should never be removed or replaced without first switching off the supply.

Your responsibility

When working on a particular circuit or machine, you must be certain that the supply cannot be switched on without your knowledge. You must satisfy yourself that the circuit is DEAD and SECURED in the OFF or OPEN position with a WARNING sign displayed at the POINT of ISOLATION.

Always check that the circuit is dead, using an approved voltage indicator or test lamp, to prevent danger from electric shock (see Safe Isolation Procedure on the next page).

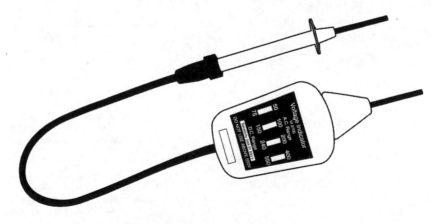

An approved voltage indicator

41

Safe isolation procedure

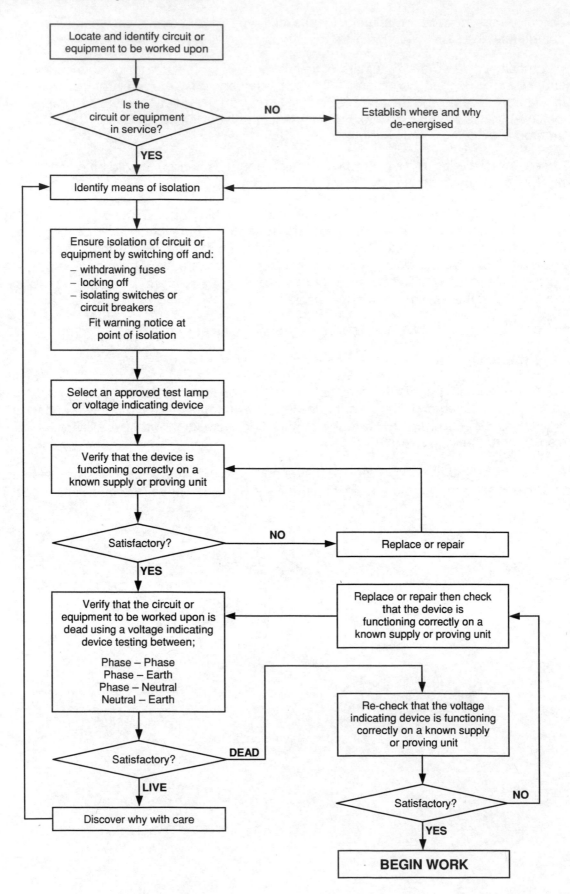

Test equipment

Test equipment for use by electricians should comply with, and be used in accordance with, the Health and Safety Executive (HSE) Guidance Note: General Series GS38.

Neon (screwdriver) testers should not be relied on since the neon will not indicate supplies at low potentials. There is also the risk of receiving an electric shock if the resistor breaks down.

Flex and lampholder (homemade) test lamps should never be used. These are extremely dangerous because:

- mains potential is present in all the components

- the flex and bulb are vulnerable to damage

- there is usually no fuse in circuit.

Approved test lamps have a robust plastic body to contain the circuitry. The circuit, in turn, is fitted with either resistors to limit the potential or resistors and fuse(s) (500 mA) to give complete protection in case of any fault occurring.

Voltage indicators or test lamps do not fail safe, they FAIL UNSAFE, and must be proved IMMEDIATELY BEFORE and IMMEDIATELY AFTER use on a known supply or proving unit of a voltage equal to that under test.

Local isolation

It has been the practice of British Gas to ask for the electrical supply point that will be used to supply a central heating system to be installed as close as practical to the boiler and in a readily accessible position. This type of supply point is a 13 A BS 1363 un-switched socket outlet and plug-top. This combination provides local isolation.

A more acceptable method of providing local isolation is to install a double-pole switch fuse spur unit, in which the fuse carrier can be withdrawn but not removed and a small padlock fitted, as illustrated below.

When isolating a central heating unit, it is important to remember that the system will be electrically connected to remote components within the property. Therefore, to ensure total isolation, all terminals at the appliance or wiring centre must be tested for the absence of electricity, with all the controls calling for heat. This will ensure, for example, that any temperature-controlled zoning system that has been wired from a separate power source will be identified and can be made safe before work is carried out on the system.

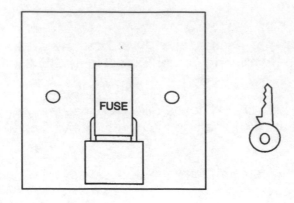

Isolation and switching

The IEE regulations state that a means of isolation is required for circuits and equipment so that a skilled person can carry out work on that installation or equipment safely.

Devices for isolation should have their contacts or other means of isolation externally visible or clearly and reliably indicated when in the **OFF** OR **OPEN** POSITION.

Each device for isolation should be clearly identified by position or durable marking to indicate the installation circuit or equipment it isolates.

In a domestic or similar installation, the main switch in a consumer unit must be double pole.

Functional switching

The IEE regulations also require a means of interrupting the supply on load for every circuit (for example, a switch for a lighting point or immersion heater). Remember that, in bathrooms, insulated cord-operated switches should be used. An insulated pull cord is permitted in Zones 1 and 2, but the switch body must be in Zone 3 (see page 85).

3. INSTRUMENTATION

Measurement of voltage, current and resistance

The multimeter

Possibly the most widely used measuring instrument is the 'multimeter'. This instrument is sometimes referred to as an AVO because it can measure **A**mps, **V**olts and **O**hms.

When used for different purposes, the multimeter has very different characteristics, and it is a mistake to assume that this unit remains the same whichever function it is being used for. This can also be a recipe for disaster. There is, therefore, clearly a need for very special caution when using a multimeter.

You will meet many different makes and models in your work, but they usually fall into one of two categories – analogue or digital.

The analogue meter uses a moving needle against a scale that has to be interpreted, like a watch with hands. The digital meter displays a read-out in numbers – the same as a digital watch.

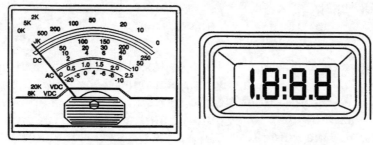

Analogue display   Digital display

Selection of correct function and range

By turning the selector switch you can choose the electrical unit and range you wish to measure:

DC voltage	AC voltage	Resistance (Ω)
DC current	AC current	

Remember, whenever you have finished using the meter, always turn the selector switch to the off position.

Instrument preparation

Before the instrument is used, you should check that it is calibrated by reading the calibration label that is attached to the instrument. The battery state should be confirmed and the instrument test leads should be checked to make sure they are in good order and fit for purpose. For measuring voltage, fused test leads should be used in accordance with HSE GS38.

When an arc is created, this ionises the surrounding air, introducing further flashovers that can quickly engulf the working area and that can prove fatal.

Electric test equipment should be:

- properly constructed

- properly maintained

- used in a way that prevents danger.

Test leads and probes:

- test probes should have finger barriers

- the probes should be insulated with an exposed metal tip of a maximum of 4 mm (ideally not more than 2 mm) across any surface, or spring-loaded screened probes should be used

- fused test leads should be used when measuring voltage (500 mA HRC fuses are usually contained within the probes)

- the test leads should be adequately insulated and suitably flexible

- the should be coloured so that one lead can be identified from the other.

Causes of accidents:

- inadequately insulated test probes (with excessive bare metal)

- excessive current flowing through test probes, leads and instruments due to the instrument being set incorrectly. **For example, if the instrument is set to current but you are attempting to measure voltage, a short circuit will be created, resulting in a large current flowing through the test leads and the instrument. This could result in a shock and serious burns**

- the incorrect use of test equipment. For example, a multimeter must not be connected to a circuit whose voltage exceeds the maximum working voltage for that instrument.

Analogue multimeters

Correct calibration

Set the range switch to the lower value of DC voltage and touch the test leads together; the needle should read zero. If the value is not zero, the accuracy of the readings will be affected. In order to adjust the reading to zero, the adjustment screw must be turned until zero is obtained.

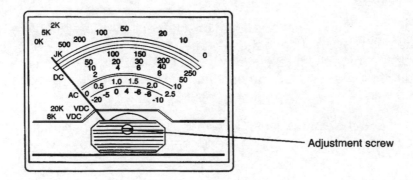

Adjustment screw

Battery condition and continuity of test leads

To check this, select the lower ohms range and bring the test probes into contact with each other. If the battery is healthy and the test leads are satisfactory, you will be able to zero the reading on the ohms scale using the adjuster. If there is no movement of the pointer, it is likely that the test leads or connections are faulty. If there is movement but you cannot zero the pointer, then the battery is suspect.

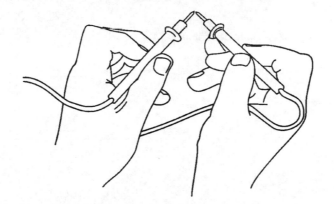

Digital multimeter

When you have confirmed that the instrument is in calibration by checking the attached label, check and fit the test leads, then confirm the instrument's battery condition is acceptable.

This type of test instrument may be 'auto ranging'. This means that, for measuring resistance, you would select the ohms (Ω) range and the instrument would then measure from zero up to its maximum value (e.g. 3 megohms (3 MΩ)).

Any test lead resistance will have to be nulled or deducted from the readings obtained.

Note: Meters and test leads must conform to HSE Guidance Note GS38.

Remember

There are some things you must do every time you use a meter. You must check that:

- the meter is in good order and calibrated

- the test leads are in a good condition and that the continuity is correct

- the battery is in good condition.

To measure the values of voltage, resistance and low values of current, always select the correct function and the correct range. If you are unsure which range to select, then start at the highest range and change progressively to a lower range to achieve a reasonable reading and to prevent overloading.

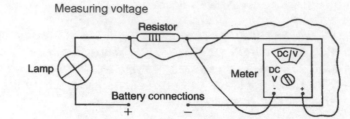

Measuring voltage

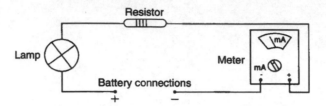

Measuring current

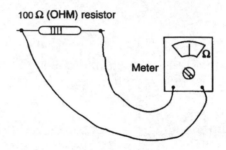

Meauring resistance

100 Ω (OHM) resistor

Remember

When you are using any meter to measure resistance, remember the following so that you obtain accurate readings and prevent damage to the meter:

- never connect the meter into a circuit with voltage present or current flowing

- always make sure the component or circuit you are checking is isolated from any other components or circuit.

Avoid read errors

Avoid errors on analogue meters that are caused by the incorrect position of the meter in relation to your eyes. This error is called parallax error. To avoid parallax error, you should ensure all readings are taken with your eyes directly above the pointer. If a mirror is fitted to the scale, the pointer's reflection should disappear.

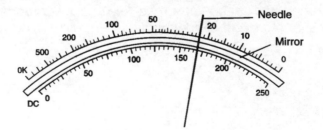

Instrument safety

- always ensure that the instrument is set correctly to the required function and range

- for voltage measurements, never switch to current or resistance ranges

- when measuring voltage, use fused test leads

- exercise great care when measuring voltages greater than 50 V

- for measuring resistance or, if using a buzzer/diode test function, the circuit or equipment under test must be de-energised

- instrument test leads must be in good order and fit for the purpose, as in HSE GS38

- do not use the instrument on voltages or currents in excess of the instrument's capability.

Measuring larger values of current

When commissioning an electrical installation or when working on refrigeration systems, it is often necessary to check the actual load of individual circuits or compressors to verify the design values. To do this involves disconnecting the circuits and inserting the ammeter test leads in series with the loads to determine the current drawn from the supply. An alternative and much simpler method is to use a clamp-type instrument, as shown below. This uses the principle of the current transformer, enabling the current to be measured without disconnecting the circuit conductors.

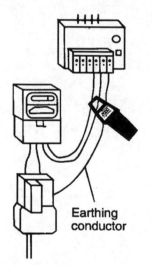

Earthing
conductor

4. ELECTRICAL REGULATIONS

The regulations and standards that cover electrical installations and practices associated with plumbing, gas, heating and ventilation, and refrigeration installations can be broken down into categories, as follows.

Statutory

The Health and Safety at Work Act 1974
The Electricity Safety, Quality and Continuity Regulations 2002
The Electrical Equipment (Safety) Regulations 1994
The Electricity at Work Regulations 1989
The Building Standards (Scotland) Regulations 1990
The Provision and Use of Work Equipment Regulations 1998
The Building Regulations 2000 for England and Wales Approved Document P

British Standard and Codes of Practice

BS 7671 Requirements for Electrical Installations (The IEE Wiring Regulations) – see Appendix 1.

Appendix 2 Memorandum of Guidance on the Electricity at Work Regulations 1989.

HSE Guidance Notes

Appendix 1 Memorandum of Guidance on the Electricity at Work Regulations 1989.

The Electricity Safety, Quality and Continuity Regulations 2002

These regulations impose requirements with regard to the installation and use of electric lines and apparatus of the distributors of electricity, including provisions for connections with earth.

The regulations also specify that the voltage at the supply terminals shall be no greater than 10% above or 6% below the declared voltage of 230 V for single-phase supplies or 400 V for three-phase supplies.

The Electrical Equipment (Safety) Regulations 1994

These regulations apply to AC equipment operating at voltages above 50 V and below 1,000 V. They also apply to DC equipment operating above 75 V and below 1,500 V.

Electrical equipment, together with its component parts, should be made in such a way as to ensure that it can be safely and properly assembled and constructed.

The equipment should be designed and manufactured to ensure protection against hazards, providing that the equipment is used in applications for which it was made and that it is adequately maintained.

These regulations cover domestic electrical equipment and electrical equipment used in the workplace.

The regulations are intended to establish a single market in safe electrical equipment, and they provide a high level of protection for consumers throughout the EEC. In the UK these regulations also cover second-hand equipment.

All new electrical equipment that is intended for supply in the UK since 9 January 1995 must comply fully with the requirements of these regulations and, as such, carry the CE marking.

The CE marking is a visible declaration by a manufacturer or their representative that the electrical equipment concerned fully meets the requirements of these regulations.

The Provision and Use of Work Equipment Regulations (PUWER) 1998

These regulations require employers to ensure that all work equipment is suitable for its purpose, is properly maintained and that appropriate training is given.

Approved Document P

Electrical Safety Approved Document P is a new part of the Building Regulations 2000 for England and Wales and is effective from 1 January 2005.

The purpose of Part P (as it is commonly known) is to improve the standard of competence of electrical installers and to try to reduce the numbers of deaths, injuries and fires caused by defective electrical installations.

All electrical work in dwellings (this includes outbuildings, garages, sheds, greenhouses, gardens, common or shared amenity areas of flats, etc.) will need to be notified to the building control department of the local authority before work commences, unless:

- the work is carried out by a prescribed competent person (this individual or company must be authorised to self-certify their work)

- the proposed work is minor work and is not in a kitchen or bathroom (or other area classed as a special location by BS 7671 IEE Wiring Regulations).

Minor work means the replacing of electrical accessories (e.g. sockets, switches, etc.). It also covers adding sockets or lights, etc., to an existing circuit. **It does not cover the installation of a new circuit.**

All electrical work (including minor work) in kitchens and special locations, e.g.:

- bath or shower rooms

- swimming pools

- saunas

- garden lighting and power

will need to be notified to building control or be self-certified by an authorised competent person.

The Electricity at Work Regulations 1989

Introduction

These regulations came into force on 1 April 1990. They have been written to reduce the increasing numbers of accidents involving electricity.

Their purpose is to require precautions to be taken against the risk of death or injury from electricity in work-related activities. The emphasis is on the prevention of danger from electric shock, burns, electrical explosion or arcing, or from fire or explosion initiated by electrical energy.

The regulations apply wherever the Health and Safety at Work Act applies, wherever electricity may be encountered and to all persons at work. The areas covered are those associated with the generation, provision, transmission, transformation, rectification, conversion, conduction, distribution, control, storage, measurement or use of electrical energy – everything from a 400 kV overhead line to a battery-powered torch. There are no voltage limits to these regulations.

In order to provide guidance on the interpretation of the regulations, the Health and Safety Executive (HSE) has produced a memorandum of guidance to assist persons involved in the design, construction, operation or maintenance of electrical systems and equipment. Therefore the assessment of danger, and how the regulations are to be applied to overcome it, will be the continuous responsibility of both the employer and employee, who must work as a team to achieve continual compliance with the regulations.

When persons who design, construct, operate or maintain electrical installations and equipment need advice, they should refer to guidance, such as may be found in codes of practice or HSE documents, or they should seek expert advice from persons who have the knowledge and experience to make the right judgements and decisions and who have the necessary skills and ability to put them into effect. It must be remembered that a little knowledge could be sufficient to make electrical equipment function, but it usually requires a much higher level of knowledge and experience to ensure its safety.

The regulations

The following is a brief summary of each of the regulations. The emphasis is on electrical systems and equipment associated with plumbing, heating and ventilating, refrigeration and gas installations.

Regulation 1: Citation and commencement

These regulations are entitled *The Electricity at Work Regulations 1989* and they came into force on 1 April 1990.

Regulation 2: Interpretation

This regulation explains the meaning of the following:

- systems
- electrical equipment
- conductors
- danger
- injury.

Regulation 3: Persons on whom duties are imposed by these regulations

It shall be the duty of every employer and self-employed person to comply with the regulations for matters within their control.

It shall be the duty of every employee when at work to:

- co-operate with their employer to enable the employer to comply with the provisions of the regulations
- comply with the regulations for matters within their control.

Regulation 4: Systems, work activities and protective equipment

This regulation has a very wide application and requires all electrical systems to be constructed and maintained in such a way to prevent danger. This includes the design of the system and the correct selection of equipment, as well as regular inspection and maintenance to ensure the continuing safety of the system and associated equipment.

The requirement for maintaining detailed and up-to-date records is also stressed.

System construction

It is the responsibility of the designer, installer and inspector to meet this obligation.

The electrical inspection and test carried out on the commissioning of an installation are to confirm the installation complies with the designer's intentions and has been constructed, inspected and tested in accordance with BS 7671.

System maintenance

It is the responsibility of the duty holder to meet this obligation.

This refers to the necessity to monitor the condition of the system throughout its life, by implementing a programme of periodic inspection and testing.

Records of maintenance (including test results) should be kept. This will allow for the monitoring the installation's condition and for the effectiveness of the programme of inspection and testing.

With regard to work activities, the overriding consideration is that work should not be carried out on a system unless it is 'dead'. The circuit or equipment to be worked on must be safely isolated (which will include locking-off, labelling, etc.) and the circuit proved dead at the point of work before work starts.

Systems with more than one possible supply (for example, extractor fans with timer circuits) will need extra care to ensure all relevant circuits are properly isolated. The test equipment used (for example, an approved voltage indicator or test lamp) to prove the circuit or equipment dead must itself be proved as functioning correctly, immediately before and after testing. This can be done using a known supply point or a voltage proving unit.

The last part of this regulation includes the requirement that protective equipment (such as special tools, protective clothing and insulating materials) must be suitable for the purpose, maintained in good condition and be properly used. Examples are insulating gloves and floor mats which are covered by British Standards (for example, BS 697 Specification for Rubber Gloves for Electrical Purposes and BS 921 Specification for Rubber Mats for Electrical Purposes). Records should be kept concerning the issue of protective equipment.

Regulation 5: Strength and capabilities of electrical equipment

All equipment must be selected so that it meets appropriate standards (for example, British Standards) and operates safely under normal and fault conditions. This is to ensure its ability to withstand thermal, electromagnetic, electrochemical or other effects of electrical current which will flow when a system is operating. Equipment must be installed and used in accordance with the instructions supplied.

Regulation 6: Adverse or hazardous environments

This regulation covers the requirement to consider the kinds of adverse conditions that could give rise to danger if equipment is not constructed or protected adequately to withstand such exposure. Therefore it is necessary, at the design stage or when assessing the maintenance requirements of equipment, to consider the effects of mechanical damage, weather and corrosive substances.

Examples of protection from mechanical damage include running cables through holes in the centre of a joist (instead of in a slot at the top of the joist) where the cable could be struck by a nail when fastening down the floorboards.

In the case of weather, when installing frost thermostats, the thermostat must be suitable for outdoor use, with an enclosure and a cable entry gland that have an appropriate IP rating.

Regulation 7: Insulation, protection and placing of conductors

This regulation requires that all conductors in a system which may give rise to danger is covered with a suitable insulating material and further protection where necessary (e.g. conduit, trunking, etc.).

The IEE Wiring Regulations provide information and advice for electrical installations up to 1,000 V AC.

Regulation 8: Earthing or other suitable precautions

This regulation applies to any conductor, other than a circuit conductor, which may become charged with electricity, either as a result of use or because of a fault occurring in the system (for example, circuit protective conductors (CPCs)).

Typical techniques are:

- earthed equipotential bonding automatic disconnection (EEBAD). Earthing enables earth faults to be detected and the supply to the faulty circuit or equipment to be cut off automatically by the operation of the circuit protective device

- connection to a common voltage reference point on the system. UK public supply systems have their transformer neutral point connected to earth. The voltage reference point is the mass of earth

- equipotential bonding. All exposed and extraneous conductive parts are inter-connected so that no dangerous potentials can exist between these parts

- double insulation

- use of reduced voltage. 110 V centre-tapped transformers with 55 V to earth, or extra-low voltage (maximum values of extra-low voltage are 50 V AC or 120 V DC)

- use of RCDs to provide supplementary protection, in addition to fuse and circuit breakers

- separated or isolating transformers (found in bathroom shaver sockets and the supply system for whirlpool baths).

Regulation 9: Integrity of referenced conductors

This refers to the maintenance of the integrity of the earth and/or neutral conductors (for example, not inserting fuses, MCBs or switches into conductors connected to earth).

Regulation 10: Connections

This regulation requires all joints and connections to be both mechanically and electrically suitable for use.

It therefore covers connections to the terminals of plugs, socket outlets, fused spur units, junction boxes and appliances. Hence, all terminations of cables and flexible cords must be such that the correct amount of insulation is removed, the connections are tight and verified by a tug on the conductor, and no stress or strain is placed on the conductor or the cable itself. This can be achieved by using cord grips in plug-tops or appliances and by clipping cables around entries into junction boxes.

Regulation 11: Means for protecting from excess of current

To meet the requirements of this regulation, the means of protection is likely to be fuses or circuit-breakers to guard against overload and fault current. The IEE regulations provide guidance on this subject. The type of electrical protection must be correctly selected, installed and maintained.

Regulation 12: Means for cutting off the supply and for isolation

This regulation covers two separate functions: cutting off the supply and isolation

Devices used for cutting off the supply are usually switches that must be capable o disconnecting the supply to equipment under normal operating or fault conditions.

They should be clearly marked to show their relationship to the equipment they contro unless that function is obvious to persons who need to operate them.

Devices used for isolation are usually switches that have the capability of establishing, when operated, an air gap with sufficient clearance distances to ensure that there is no way in which the isolation gap can fail electrically.

The position of the contacts or other means of isolation should either be externally visible or clearly and reliably indicated.

Typical isolating devices would be a double-pole switch in a consumer unit and, in the case of an appliance, a 13 A plug and socket.

Both devices (for switching and isolating) should be positioned so that there is ease of access and operation, and the area adjacent should be kept free from obstructions.

Regulation 13: Precautions for work on equipment made dead

This regulation states that the requirements for such precautions should be effective in preventing electrical equipment from becoming charged with electricity in such a way that would give rise to danger.

The procedures for making equipment dead will usually involve switching off the isolating device and locking it off. If such facilities are not available, the removal of fuses or links (which are then kept safe) is an alternative and secure arrangement if proper control procedures are in place.

Regulation 14: Working on or near live conductors

This regulation sets out the requirements for working on or near equipment which has not been isolated and proved dead.

A typical example of live work would be live testing (for example, the use of a suitable voltage indicator or multimeter on mains power).

The factors that have to be considered in deciding if it was justifiable for live work to proceed would include:

- when it is not practical to carry out the work with the conductors dead (for example, when measuring voltage).

When working on or near conductors which are live, suitable precautions would be:

- the use of people who are properly trained and competent to work safely on live equipment

- the provision of adequate information about the live conductors involved, the associated electrical system and the possible risks

- the use of suitable tools, including insulated tools, equipment and protective clothing

- the use of suitable insulated barriers or screens

- the use of suitable test instruments and test probes (GS38).

Testing to establish whether electrical conductors are live or dead should always be done on the assumption they are live until such time as they have been proved dead.

When using test instruments or voltage indicators for this purpose, they should be proved immediately before and immediately after use on a known supply or proving unit. It must be remembered that, although live testing may be justified, it does not follow that such justification can be made for the repair work to be carried out live. It should be carried out with the conductors safely isolated.

Regulation 15: Working space access and lighting

The purpose of this regulation is to ensure that sufficient space, access and adequate illumination are provided while persons are working on, at or near electrical equipment in order that they may work safely.

Regulation 16: Persons to be competent to prevent danger and injury

The object of this regulation is to ensure that persons are not placed at risk due to a lack of skills on their behalf or the behalf of others when dealing with electrical equipment and the work associated with it.

The requirements are that persons must possess sufficient technical knowledge or experience or be supervised.

For the purpose of this regulation, technical knowledge or experience may include:

- an adequate knowledge of electricity

- adequate experience of electrical work

- an adequate understanding of the system to be worked on and practical experience of that system

- an understanding of the hazards which may arise and the precautions which need to be taken during work on a system

- an ability to recognise at all times if it is safe to continue to work.

Regulation 29: Defence

This provides a defence for a duty holder who can prove they have taken all reasonable steps and exercised all due diligence to avoid committing an offence.

Regulation 30: Exemption certificates

This regulation allows the Health and Safety Executive to grant exemption certificates in special cases. This is unlikely to be required for individuals or organisations involved in electrical systems for plumbing, heating and ventilating, refrigeration and gas installations.

Regulation 31: Extension outside Great Britain

This concerns the application of the regulations outside Great Britain.

Regulation 32: Disapplication of duties

This states that the regulations do not apply to seagoing ships, hovercraft and aircraft but may apply to electrical equipment on vehicles if the equipment could possibly give rise to danger.

Regulation 33: Revocations and modifications

This regulation lists previous regulations which have been replaced or modified (including the Electricity (Factories Acts) 1908 and 1944).

Appendix 1
Lists HSE and HSC publications on electrical safety.

Appendix 2
Lists other publications having an electrical safety content.

Appendix 3
Gives information concerning working space and access, together with historical comments on revoked legislation.

5. ELECTRICAL DISTRIBUTION

History of development

In 1878, Thomas Edison developed the first electric light bulb. It was a marvellous invention, but Edison had to find a way to get electricity to the users so that they could buy and use his new light. At this time Edison was using direct current (DC) electricity, which is transmitted by two wires, one 'positive' and one 'negative'. Unfortunately, he found that the wires he was using to carry the electricity had a resistance to the current and, if the bulb was more than about 3 km away from the power station, that the light it produced was too dim to be of any use. In order to provide everyone with a sufficiently bright light, Edison would have had to build power stations every 6 km.

In 1885, Westinghouse and Stanley's experiments showed that they could reduce the energy loss by using alternating current and a transformer. However, it took Edison twenty years to concede that higher efficiencies could be achieved through AC power transmission.

The basic principles used by Westinghouse and Stanley are still used today in our National Grid. The main problem is that wires have resistance. When a current passes through a resistor, heat is generated and some of the energy from the electricity is lost through this heat. We can reduce this heat loss in one of two ways. The resistance of the wire can be reduced by making it thicker, but this would cost more and would use up the valuable resources of copper and other metals. The other solution is to reduce the amount of current flowing.

To work out the power transferred by an electric current we used the formula:

Power transferred = voltage supplied x current flowing.

Clearly, we want to keep the power transferred the same so, in order to reduce the current, we must increase the voltage. With DC electricity this is difficult, but with AC we can use a transformer.

By having more turns in the output (secondary coil) of the transformer, we can increase the voltage and therefore reduce the current. When we need to reduce the voltage, we pass the electricity through a transformer with fewer wires in the secondary coil than in the primary coil. This reduces the voltage and increases the current.

The distribution system

The power-generating companies use step-up transformers to increase the voltage on overhead cables to 275,000 V (275 kV) or, in some cases, 400,000 V (400 kV). The electricity is transported across the country to towns and cities, often by overhead cables connected to pylons. The voltage is reduced to the required level by local transformers, which are usually called substations.

In the British Isles, electricity is usually supplied to homes at 230 V, but some heavy users may receive 400 V and industries either 11 kV or 33 kV.

The local substations, like the electricity pylons, are not playgrounds. Every year young people are killed because they ignore all the warnings and venture in. Because of the amounts of energy supplied to these substations, very few people get a second chance if they touch the equipment.

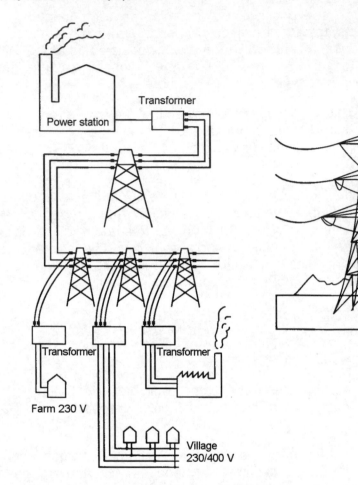

The National Grid

AC supply

The electricity we use from the power sockets in our homes and that used by industry is different from the electricity we get from batteries. The electricity from batteries is direct current (DC) whereas the mains supply is alternating current (AC). In the United Kingdom, AC electricity has a frequency of 50 Hz (50 cycles per second). This means that the live AC wire will be changing from negative to positive and back 50 times in every second.

The supply to a domestic property is 230 V AC (although some houses may have a 400 V three-phase supply).

On a new development the local electrical distributor would arrange for the properties to be fed in sequence from different phases in order to try to balance the load across all three phases of the main supply cable in the road.

In theory this means that the first house in the road is connected to the red (L1) phase and neutral, the next house to the yellow (L2) phase and neutral, and the next house to the blue (L3) phase and neutral (all the houses are connected to the same neutral conductor).

The difference between phase and neutral in a property is 230 V AC, but between the phase of one house and the phase of the house next door would be 400 V AC. Therefore it is important to keep the electrical equipment from the two properties apart in case a fault occurs which could expose a person to 400 V AC.

Where possible, this will be the method of installation adopted by the distributor but, in practice, this phase rotation will not always apply. On older installations there could be some variation and, in one-off new services, the cable might be connected to whichever phase was easier to use at the main supply cable in the road.

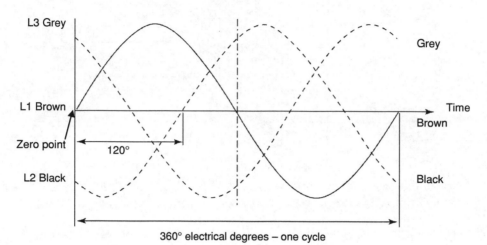

360° electrical degrees – one cycle

Each phase has a different value of voltage at any given time, and they are said to be 120° out of phase with each other.

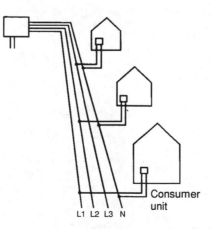

The diagram above shows a three-phase electricity supply along a street; each house's consumer unit is connected to a different phase from that of its neighbour.

Electricity into the home

The distributor's service cable to a domestic property commonly enters the property or the meter box (if fitted) from underground. The cable is terminated at the service cut-out. Occasionally, the distributor's cable is run overhead and therefore enters the property or meter box from above.

Two single-core, double-insulated cables, one for phase and one for neutral (usually 25 mm² for domestic installations), are connected from the distributor's cut-out to the meter. The meter is then connected to the distributor's double-pole switch or, if there is no switch, then directly to the consumer unit. These cables are commonly known as meter tails.

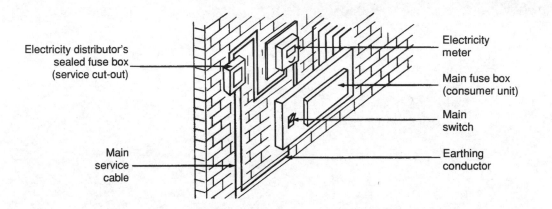

These two conductors are both live conductors. The phase conductor is coloured red (brown) and the neutral is coloured black (blue). (Harmonized colours in brackets.)

A separate green and yellow earthing conductor (usually 16 mm² for domestic installations) is connected from the earth facility provided by the distributor (or the consumer's earth electrode if a TT system) to the main earthing terminal for the installation, usually the earth bar inside the consumer unit.

6. ELECTRICAL SUPPLY SYSTEMS

An electrical system consists of a single source of electrical energy and an installation. BS 7671 (IEE Wiring Regulations) identifies types of systems as follows (depending on the relationship of the source and of the exposed conductive parts of the installation to earth).

Types of electrical system

- **TN system**
  A system having one or more points of the source of energy directly earthed, the exposed conductive parts of the installation being connected to that point by protective conductors

- **TN-C system**
  In which neutral and protective functions are combined in a single conductor through the system

- **TN-S system**
  This has separate neutral and protective conductors throughout the system

- **TN-C-S system**
  In which neutral and protective functions are combined in a single conductor in part of the system

- **TT system**
  A system having one point of the source of energy directly earthed, the exposed conductive parts of the installation being connected to earth electrodes electrically independent of the earth electrodes of the source

- **IT system**
  A system having no direct connection between live parts and earth, the exposed conductive parts of the electrical installation being earthed.

Classification of systems

A system consists of an electrical installation connected to a supply. Systems are classified with the following letter designations.

SUPPLY earthing arrangements are indicated by the first letter:

- **T** – one or more points of the supply are directly connected to earth

- **I** – supply system not earthed, or one point earthed through a fault-limiting impedance

INSTALLATION earthing arrangements are indicated by the second letter:

- **T** – exposed conductive parts connected directly to earth

- **N** – exposed conductive parts connected directly to the earthed point of the source of the electrical supply (the point where neutral normally originates).

The RELATIONSHIP between the NEUTRAL and PROTECTIVE CONDUCTORS is indicated, where appropriate, by the third and fourth letters:

S – separate neutral and protective conductors

C – neutral and protective conductors combined in a single conductor.

The types of systems are:

TN-S   TT   TN-C   TN-C-S   IT

Connections to earth

The earthing arrangement of an installation must be such that:

- the value of impedance from the consumer's main earthing terminal to the earthed point of the supply for TN systems or to earth for TT and IT systems is in accordance with the protective and functional requirements of the installation and expected to remain continuously effective

- earth fault and protective conductor currents which may occur under fault conditions can be carried without danger, particularly from thermal, thermomechanical and electromechanical stresses

- they are robust or protected from mechanical damage appropriate to the assessed conditions.

The installation should be so installed as to avoid risk of subsequent damage to any metal parts or structures through electrolysis.

Voltage ranges

Extra-low voltage (ELV)

0 V to 50 V AC (RMS) or 120 V ripple-free DC, whether between conductors or to earth.

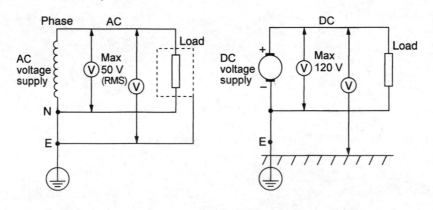

Extra-low voltage systems

66

Low voltage

Exceeding ELV, but not exceeding 1,000 V AC or 1,500 V DC between conductors or 600 V AC (RMS) or 900 V DC between conductors and earth.

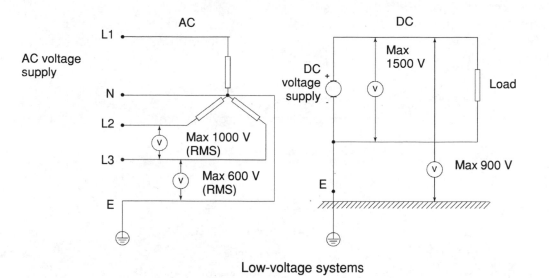

Low-voltage systems

System earthing arrangements

TN-S systems

This is likely to be the type of system used where the electricity distributor's installation is fed from underground cables with metal sheaths and armour.

In TN-S systems, the consumer's earthing terminal is connected by the distributor to their protective conductor (i.e. the metal sheath and armour of the underground cable network), which provides a continuous path back to the star point of the supply transformer and which is effectively connected to earth.

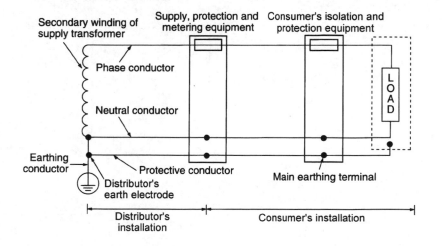

TT systems

This is likely to be the installation used if the distributor's installation is fed from overhead cables and where no earth terminal is supplied.

With such systems, the earth electrode for connecting the circuit protective conductors to earth has often to be provided by the consumer. Effective earth connection is frequently difficult to obtain and, in such cases, a residual current device should be installed in addition to an overcurrent protective device.

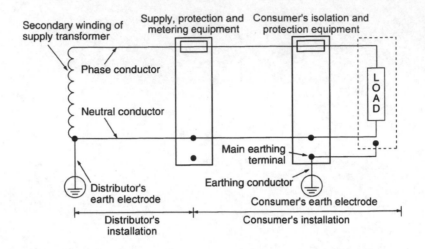

TN-C-S systems

When the distributor's installation uses a combined protective and neutral (PEN) conductor, this is known as a TN-C supply system.

Where consumer's installations consisting of separate neutral and earth (TN-S) are connected to the TN-C supply system, this combination is called a TN-C-S system. This is the system usually provided to the majority of new installations, and it is referred to as a PME system by the distributor.

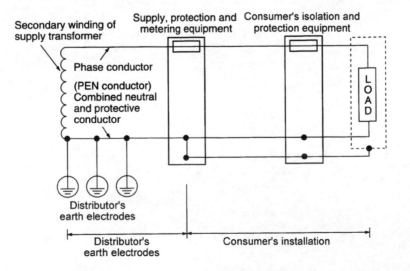

TN-C systems

Where a combined neutral and earth conductor (PEN conductor) is used in both the supply system and the consumer's installation, this is referred to as a TN-C system.

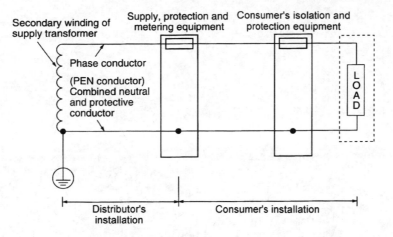

Regulation 8 (4) of the Electricity Safety, Quality and Continuity Regulations 2002 prohibits a consumer from combining the neutral and protective functions in a single conductor within the consumer's installation.

Earthing arrangements and terminations

The diagrams below show what the three most common earthing arrangements look like in practice.

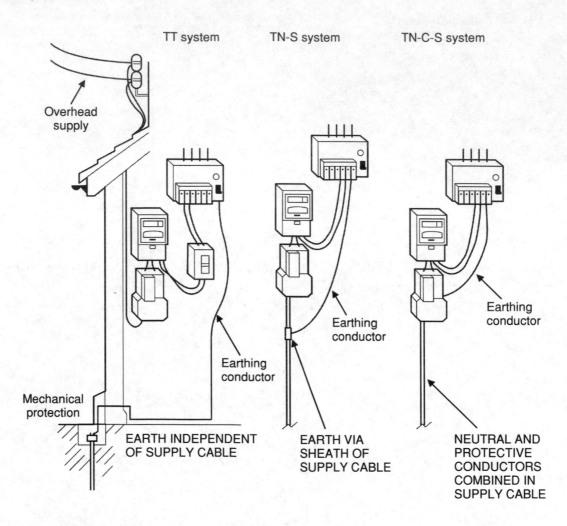

In the United Kingdom, distributors have to comply with the Electricity Safety, Quality and Continuity Regulations 2002.

Full discussions with the relevant distributor are essential when planning or installing a customer's installation in order to obtain the specifications for any special requirements (e.g. the size of the earthing conductor and the main bonding conductors, the values of earth loop impedance (Z_e) and the prospective fault current for the electrical supply).

7. CONTROL AND PROTECTION EQUIPMENT

The distributor's equipment

Most of our electricity is generated at large, remote power stations and then transmitted the relatively long distances into our towns and cities. At a certain point it becomes the responsibility of the local distributors, who distribute it to our premises through underground and service cables. It is the service cable that is terminated in a mutually agreed position within the premises.

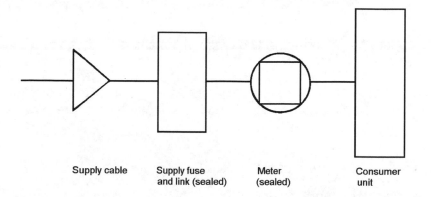

| Supply cable | Supply fuse and link (sealed) | Meter (sealed) | Consumer unit |

At the service termination point, the service cable is connected to a service fuse (normally 80 A or 100 A) that is designed to protect the distributor's cables and equipment in the event of a fault occurring on the consumer's premises.

The distributor installs connections from the service fuse to the meter, which records the amount of electricity consumed in units or kilowatt hours (kWh). The service fuse and meter are sealed by the distributor – consumers or contractors are not permitted to interfere with this part of the installation.

The consumer's installation

The consumer's installation begins at the output terminals of the distributor's meter or, in the case of a domestic installation, the distributor's meter or double-pole switch (the switch, if fitted, is on the outgoing side of the meter).

Main switches must be provided so that the installation may be isolated from the supply when alterations or extensions are made.

Consumer control units

For domestic installations where the load does not exceed 100 A, a consumer control unit is normally installed.

A consumer control unit constructed with an open back is not acceptable since the enclosure must provide protection to at least IP2X or IPXXB, with the top of the enclosure being to IP4X.

Consumer control units are manufactured to BS EN 60439-3 and usually consist of a 60, 80 or 100 A double-pole switch or a residual current device (RCD) and a number of output ways or modules.

An arrangement with eight output ways could be used to supply the following circuits:

- two 32 A ways for two-ring final circuits
- one 32 A way for the cooker circuit
- one 20 A way for the garage power circuit
- one16 A way for the immersion heater circuit
- two 6 A ways for the lighting circuits
- one spare way.

There will be many variations – no two types of installation will have exactly the same requirements.

Modern consumer control units are very versatile – manufacturers produce a wide range of components that can be fitted into the selected enclosure. These include:

- circuit-breakers
- combined circuit-breaker/RCD (RCBO)
- fuses (cartridge type)
- time switches
- contactors
- transformers.

Enclosures are selected for a given module size subject to the needs of the installation.

The main switch could be a double-pole switch or an RCD, then a range of circuit-breakers or BS 1361 fuses to protect the individual circuits.

Typically, the main switch or RCD would occupy two module ways each, and the individual circuit protective devices occupy one module each. Some control devices will take up two or even three module ways (such as contactors and time switches) and allowances should be made for these at the design stage.

Split-load consumer control units

Split-load consumer control units are the popular choice for modern domestic installations. They give sensitive RCD protection to selected circuits, with the unit being specifically designed to suit the installation requirements.

The main switch could be:

- a double-pole switch
- a 100 mA RCD
- a time-delayed RCD.

This would be the consumer's main switch for the installation and would supply a number of output ways for selected circuits that do not require 30 mA RCD protection (such as lighting). It would also supply the 30 mA RCD that protects a further number of output ways for circuits that do require a high level of earth fault protection (such as power).

Therefore, if an earth fault occurs on the ring circuit, the 30 mA RCD would operate, leaving the main switch side of the consumer unit energised and the lighting would remain on. This avoids plunging the building into darkness, with the additional risk of physical injury.

With a standard consumer unit that has a double-pole main switch, it may be possible to install an RCBO (combined RCD and circuit breaker) within the unit to give RCD protection to an individual circuit.

Both metal-clad or insulated consumer control units are available. If a conduit system is being installed, a metal-clad unit is normally used. For sheathed cable systems, an insulated enclosure is more usual.

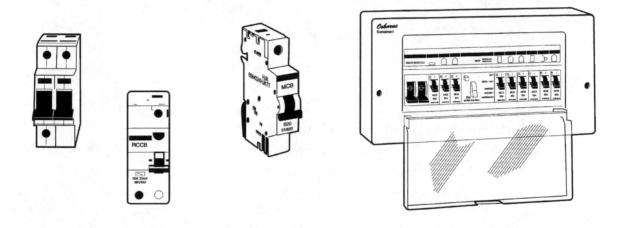

Main switchgear

Every installation must be controlled by one or more main switches. The main switchgear may consist of a switchfuse or a separate switch and fuses, and must be readily accessible to the consumer and as near as possible to the distributor's intake. The Electricity at Work Regulations state that 'suitable means shall be available for cutting off the supply and isolating the supply to any electrical equipment'.

The IEE Wiring Regulations BS 7671 require that 'effective means, suitably placed for ready operation, shall be provided so that all voltage may be cut off from every installation, from every circuit thereof and from all equipment as may be necessary to prevent or remove danger'.

Identification notices

Switchgear and control gear protective devices

Switchgear and control gear in an installation should be labelled to indicate its use. All protective devices in an installation should be arranged and identified so that their respective circuit may be easily recognised.

The following information should be made available.

- Diagrams, charts or tables indicating:
    - the type of circuits
    - the number of points installed
    - the number and size of conductor
    - the type of wiring system.

- Details of the characteristics of the protection devices for automatic disconnection, and a description of the method used for protection against indirect contact.

- The location and types of devices used for:
    - protection
    - isolation and switching.

- Details of circuits or equipment sensitive or vulnerable to tests – e.g. equipment with voltage-sensitive electronics, such as central heating controls with electronic timers and displays, and also intruder alarm equipment, certain types of RCD, etc.

Note: Information may be given in a schedule for simple installations. (See the example on the next page for a domestic installation.) A durable copy of the schedule relating to the distribution board must be provided inside or adjacent to the distribution board.

Schedule of installation at...

Type of circuit	Points served	Phase conductor mm²	Protective conductor mm²	Protective devices	Type of wiring
Lighting	10 downstairs	1 mm²	1 mm²	6 A Type B circuit-breaker	PVC/PVC
Lighting	8 upstairs	1 mm²	1 mm²	6 A Type B circuit-breaker	PVC/PVC
Immersion heater	Landing	2.5 mm²	1.5 mm²	16 A Type B circuit-breaker	PVC/PVC
Ring	10 downstairs	2.5 mm²	1.5 mm²	32 A Type B circuit-breaker	PVC/PVC
Ring	8 upstairs	2.5 mm²	1.5 mm²	32 A Type B circuit-breaker	PVC/PVC
Shower	Bathroom	10 mm²	4 mm²	40 A Type B circuit-breaker	PVC/PVC

8. EARTHING ARRANGEMENTS AND PROTECTIVE CONDUCTORS

Earthing arrangements

Purpose of earthing

By connecting the non-current-carrying metalwork to earth, a path is provided for leakage current which can be detected and, if necessary, interrupted by the following devices:

- fuses

- circuit-breakers

- residual current devices (RCDs).

General

The earth can be considered to be a large conductor which is at zero potential. The purpose of earthing is to connect all metalwork (other than that which is intended to carry current) to earth so that dangerous potential differences cannot exist either between different metal parts or between metal parts and earth.

Every means of earthing and protective conductor must satisfy the requirements of BS 7671.

Connections to earth

The earthing arrangement of an installation must be such that:

- the value of impedance from the consumer's main earthing terminal to the earthed point of the supply (TN systems) or to earth (TT and IT systems) complies with the protective and functional requirements of the installation and is expected to remain continuously effective, AND

- earth fault and protective conductor currents which may occur under fault conditions can be carried without danger, particularly from thermal, thermomechanical and electromechanical stresses, AND

- they are sufficiently robust or protected from mechanical damage appropriate to the assessed conditions.

The installation should be so installed as to avoid risk of subsequent damage to any metal parts or structures through electrolysis.

Main earthing terminals or bars

A main earthing terminal must be provided in every installation to enable the earthing conductor to connect to:

- circuit protective conductors

- main bonding conductors.

Provision must be made for disconnection of the earthing conductor for test measurement of the earthing arrangements.

The method of disconnecting the earthing terminal from the means of earthing must be such that it can only be effected with the use of tools, and may conveniently be combined within the main earthing terminal.

Protective conductors

Cross-sectional areas

The minimum cross-sectional area of protective conductors can be obtained by using Table 54G (IEE regulations). This establishes the minimum cross-sectional area of the protective conductor in relation to the cross-sectional area of the associated phase conductor. A typical example:

Cross-sectional area of phase conductor SIZE (S) mm^2	Minimum cross-sectional area of the protective conductor if it is of the same material as the phase conductor SIZE (S) mm^2
$S \leq 16$	S
$16 < S \leq 35$	16
$S > 35$	$\dfrac{S}{2}$

The minimum cross-sectional area of circuit protective conductors can also be obtained from the tables in the *On-site Guide*.

If the protective conductor does not form part of a cable, is not a conduit, ducting or trunking, and is not contained in an enclosure formed by the wiring system, the cross-sectional area should not be less than:

- 2.5 mm^2 if sheathed, or otherwise provided with mechanical protection

- 4 mm^2 where mechanical protection is not provided.

Types of protective conductor

- PVC insulated single-core cable manufactured to BS 6004 (colour green/yellow)

- PVC insulated and sheathed cable with an integral protective conductor manufactured to BS 6004

- copper strip

- metal conduit ⎤

- metal trunking systems ⎬— enclosures

- metal ducting ⎦

- MICC – cable sheath

- lead-covered cable sheath

- SWA cable armourings

- an extraneous conductive part.

When the protective conductor is formed by a wiring system such as conduit, trunking, MICC, armoured cables or sheathed and insulated cables, a separate protective conductor must be installed from the earthing terminal of the socket outlets to the earthing terminal of the associated box or enclosure.

The circuit protective conductor of final ring circuits which are not formed by the metal covering or enclosures of a cable should be installed in the form of a ring that has both ends connected to the earth terminal at the origin of the circuit (for example, the distribution board or consumer's unit).

Note: Flexible conduit must not be used as a protective conductor; an additional circuit protective conductor must always be installed within the conduit but it must be accessible at the terminations.

Main equipment bonding conductors

Main bonding conductors should connect the following to the installation's main earthing terminal (MET):

- water service pipes

- gas installation pipes

- other service pipes and ducting

- central heating and air-conditioning systems

- exposed metallic structural parts of the building.

The use of plastic pipework for installations within a building can affect equipotential bonding. Therefore:

- if the incoming service pipes are plastic, and the pipes within the installation are also plastic, they do not require main bonding

- if the incoming service pipes are plastic, but the pipes within the installation are metal, the main equipotential bonding **must** be carried out to the metal installation pipes.

The main bonding conductors should be not less than half the cross-sectional area of the earthing conductor, the minimum size being 6 mm^2 with a maximum size of 25 mm^2.

Where PME conditions apply, the cross-sectional area should be in accordance with Table 54H (see below), which gives the minimum cross-sectional area of main bonding conductors in relation to the neutral conductor of the incoming supply. The conductor sizes are for copper conductors or other conductors affording equivalent conductance.

Supply neutral conductor cross-sectional area	Main equipotential bonding conductor cross-sectional area
35 mm^2 or less	10 mm^2
Over 35 mm^2 up to 50 mm^2	16 mm^2
Over 50 mm^2 up to 95 mm^2	25 mm^2
Over 95 mm^2 up to 150 mm^2	35 mm^2
Over 150 mm^2	50 mm^2

Note: The local distributor's network conditions may require larger conductors.

The extraneous conductive parts within an installation, such as gas and water services, must be at the same potential as the exposed conductive parts (i.e. the metalwork of the electrical installation). This creates an earthed equipotential zone.

This is achieved by connecting all the exposed conductive parts of the electrical installation to the main earthing terminal by the circuit protective conductors (CPCs) and by installing main bonding conductors from the main earthing terminal to the gas, water and other services at their point of entry to the premises, as illustrated below.

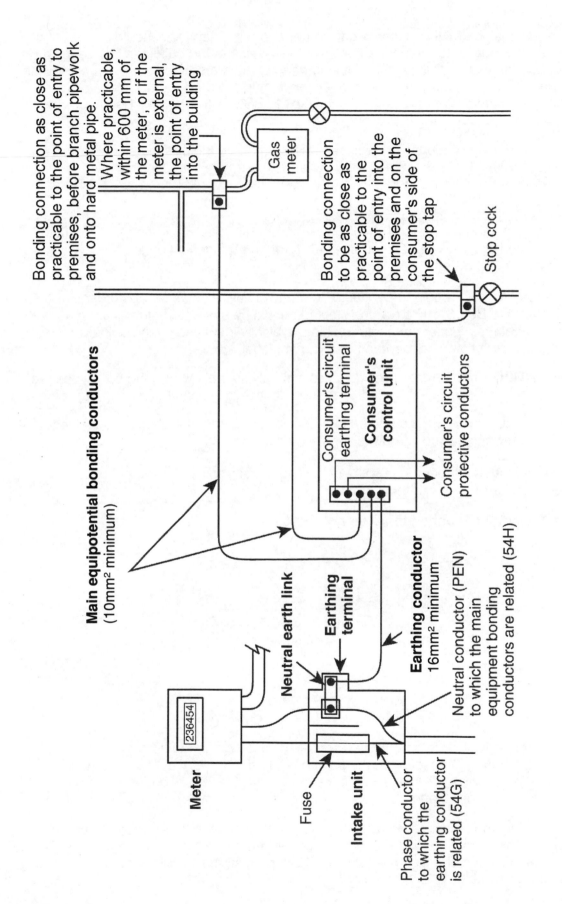

Bonding connection as close as practicable to the point of entry to premises, before branch pipework and onto hard metal pipe.

Where practicable, within 600 mm of the meter, or if the meter is external, the point of entry into the building

Gas meter

Bonding connection to be as close as practicable to the point of entry into the premises and on the consumer's side of the stop tap

Stop cock

Main equipotential bonding conductors
(10mm² minimum)

Consumer's circuit earthing terminal

Consumer's control unit

Consumer's circuit protective conductors

Neutral earth link

Earthing terminal

Earthing conductor
16mm² minimum

Neutral conductor (PEN) to which the main equipment bonding conductors are related (54H)

Meter

236454

Fuse

Intake unit

Phase conductor to which the earthing conductor is related (54G)

81

The main equipotential bonding to gas, water and other services should be made as close as practicable to the point of entry of the service to the building. The bonding should be applied on the consumer's side of any meter, stopcock or insulating section and before branch pipework. The connection is made to hard metal pipes, not to soft metal or flexible pipework, using earth clamps to BS 951 that will not be affected by corrosion at the point of connection.

If there is a meter, the bonding connection should be on the consumer's side, preferably within 600 mm of the meter outlet before any branch pipework. If the meter is external to the building then, ideally, the connection should be made as close as practical to the point of entry of the service to the premises.

The main bonding conductors should be separate (as shown), or a single conductor may be used provided it passes **unbroken** through the connection at one service and directly on to the other.

The connection of a bonding conductor to metal pipework is usually by means of an earth clamp to BS 951. The clamp selected should be suitable for the environmental conditions at the point of connection. This is typically identified by the colour of the clamp body or a coloured stripe on the warning label, where:

- a RED stripe on the label indicates it is only suitable where the conditions are non-corrosive, clean and dry (as a guide, hot pipes only)

- a BLUE or GREEN stripe on the label indicates that the clamp is suitable for all conditions (including corrosive and humid).

The clamp must be adequate for the size of conductor being connected and it must have a label attached with the words:

'Safety electrical connection – do not remove'

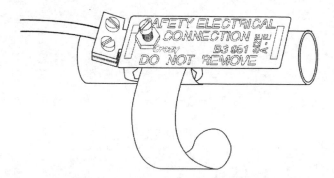

If a gas meter is fitted in an external, semi-concealed box and it is not practicable to provide a bonding connection at the point of entry due to the gas installation pipe entering the premises at low level (i.e. below the floorboards or buried in a concrete floor), then it is possible to make the bonding connection within the meter box itself.

The bonding conductor, when returning inside the premises, must pass through a separate hole above the damp course (with the hole sealed on both sides). The Gas Safety (Installation and Use) Regulations 1994 prohibit the bonding conductor passing through the same hole as a gas pipe.

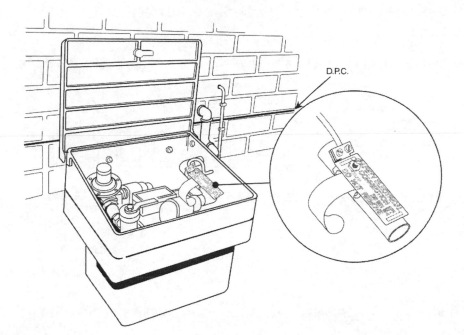

D.P.C.

A connection made to the gas installation pipe externally between the meter box and entry point to the building would encourage the risk of corrosion and mechanical damage.

If there is a water meter on the consumer's side of, and close to, the stop tap, then the main bonding conductor could be connected directly after the meter. Otherwise, the main bonding connection should be as close as possible to the point of entry on the consumer's side of the stop tap as normal.

If a water meter is installed within the premises on the consumer's side of, and some distance from, the stop tap, then the water meter should be bridged by a bonding conductor. This is to prevent damage to the working parts of the meter in the event of a fault current flowing through that section of pipework in which the meter is connected and to maintain continuity if the meter body is made from insulating material.

Typical earthing and bonding conductor arrangements for domestic installations

TN-S earthed to armour or metal sheath of the distributor's cable:

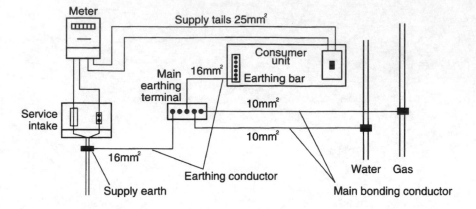

TN-C-S earthed using combined neutral and earth conductor of the distributor's cable:

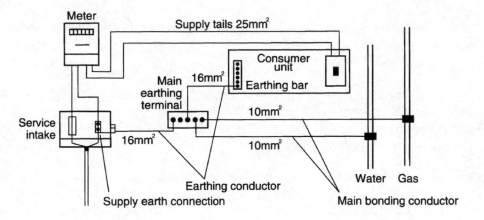

TT earthed via an earth electrode:

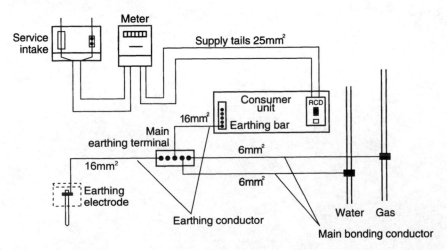

Main bonding conductors can be installed separately or as an unbroken loop.

Note: The local distributor's network conditions may require larger conductors.

Supplementary bonding conductors

A supplementary bonding conductor used to connect exposed conductive parts must have a cross-sectional area not less than the smallest protective conductor connected to the exposed conductive parts, subject to a minimum of:

- 2.5 mm² if sheathed or mechanically protected

- 4 mm² if mechanical protection is not provided.

In situations where a supplementary bonding conductor connects two extraneous conductive parts, neither of which are connected to an exposed conductive part, the minimum cross-sectional area of the supplementary bonding conductor shall be:

- 2.5 mm² if sheathed or mechanically protected

- 4 mm² if mechanical protection is not provided.

Supplementary bonding in areas of increased shock risk

In domestic premises, the areas identified as having increased shock risks are usually bath or shower rooms and the areas around swimming pools.

Rooms containing a bath or shower

This section covers baths, showers (and any associated cubicles) and their surroundings. It does not apply to emergency facilities in laboratories or industrial areas.

Areas that contain baths or showers for medical treatment or disabled persons may require special consideration.

Classification of zones

The following requirements are based on the dimensions of four zones (see the diagrams).

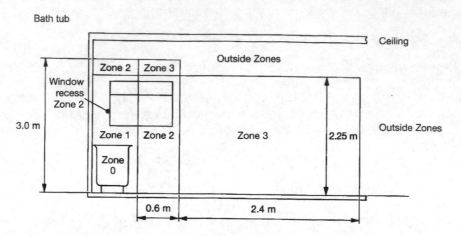

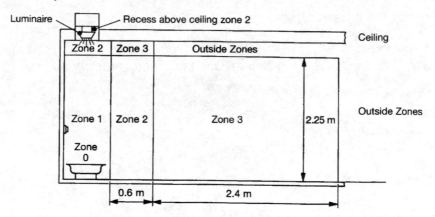

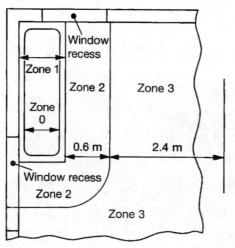

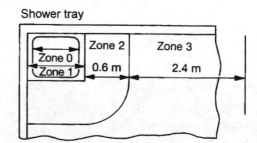

In the diagrams on the previous page, Zone 0 is the interior of the bath tub or shower tray.

The area below the bath or shower tray is classed as Zone 1 if access is without the use of a tool. If access is only by the use of a tool, then this space is considered to be outside the zones.

Electric shock protection

Where SELV or PELV is used, direct contact protection shall be provided by either:

- barriers or enclosures to IP2X or IPXXB, or

- insulation able to withstand a type test voltage of 500 V AC for one minute.

Supplementary bonding

Local supplementary bonding should be used to connect the terminal of the protective conductor of circuits supplying Class I and Class II equipment in Zones 1, 2 and 3, and extraneous conductive parts in those zones, including:

- metal water, gas and other service pipes, and metal waste pipes

- metal central heating pipes and air-conditioning

- accessible metal structural parts of the building (metal door and window frames, etc., are not treated as extraneous conductive parts unless they are connected to structural metal parts of the building)

- metal baths and shower trays.

An example of supplementary bonding in a bathroom with **metal** pipework is shown below.

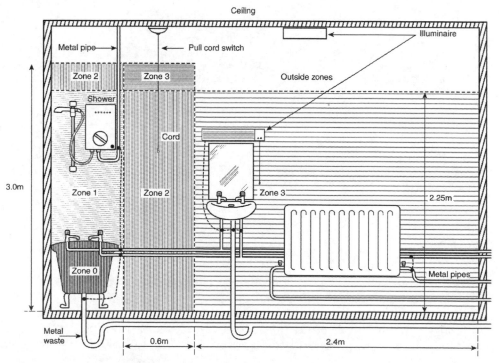

This bonding may be provided in close proximity to the location (e.g. in an airing cupboard adjoining the bathroom).

The requirements for supplementary bonding apply to Zone 1 and 2 of a room other than a bath or shower room (e.g. a bedroom where a shower cubicle is installed).

Supplementary bonding should be provided by:

- a conductor, or

- a conductive part of a permanent and reliable nature (e.g. metal pipework with soldered or compression-type fittings that are electrically continuous), or

- a combination of these.

If using metal pipework that has push-fit-type fittings (which do not provide electrical continuity) as a supplementary bonding conductor, the joints in the pipework will need to be bridged with a bonding conductor (see below).

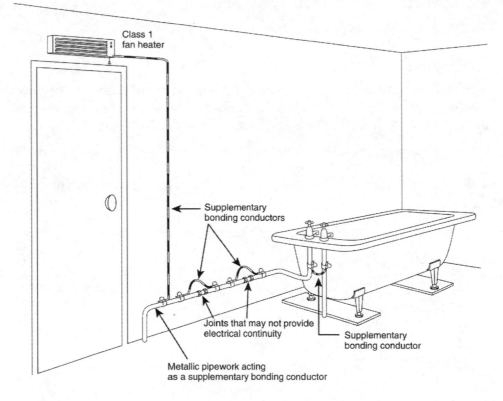

Plastic pipework

The introduction of plastic pipework and push-fit-type fittings for use on metal pipe has had an effect on supplementary bonding requirements, particularly in bathrooms.

For plastic pipe installations, there is no need to supplementary bond metal fittings supplied by plastic pipe (e.g. hot and cold taps or radiators supplied by plastic pipes).

A metal bath **not** connected to any extraneous parts (e.g. structural steelwork and supplied by plastic hot and cold pipework and with a plastic waste pipe) does not require supplementary bonding.

The use of plastic pipe can create a safer electrical installation. However, installing supplementary bonding to equipment that is connected to plastic pipework could reduce levels of electrical safety, not increase them.

An example of supplementary bonding in a bathroom with **plastic** pipework is shown below:

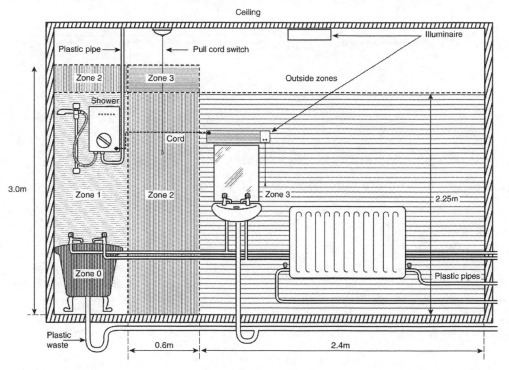

Supplementary bonding in other locations

There are no specific requirements in BS 7671 to carry out supplementary bonding in such areas as domestic kitchens, toilets and cloakrooms that may contain such items as sinks, metal pipes and washbasins (BUT NOT A BATH OR SHOWER).

Supplementary bonding can be found at sinks and/or washbasins that are installed in areas other than those that contain a bath or shower. This could be the result of previous editions of the IEE regulations that required sinks to be bonded where extraneous conductive parts were not reliably connected to the main bonding. Therefore, the bonding was applied whether required or not.

Note: Metal waste pipes in contact with earth should be main bonded to the main earthing terminal.

9. TERMINATING CABLES AND FLEXIBLE CORDS

The entry of a cable end into an accessory is known as a termination. In the case of a stranded conductor, the strands should be twisted together with pliers before terminating.

Where possible, any single conductors should be folded to ensure an effective connection. Care must be taken not to damage the conductors.

The IEE regulations require that a cable termination of any kind should securely anchor all the strands of the conductor and not impose any appreciable mechanical stress on the terminal or socket or any undue strain on the conductor itself.

A termination under mechanical stress is liable to loosen or disconnect. When current is flowing, a certain amount of heat is developed, and the consequent expansion and contraction may be sufficient to allow a conductor under stress, particularly one under tension, to loosen or be pulled out of the terminal or socket.

Unless the equipment manufacturer's instructions state otherwise, all conductors should preferably be of sufficient length to allow them to be terminated at least one more time.

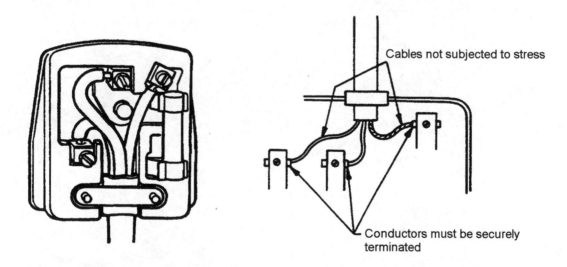

Cables not subjected to stress

Conductors must be securely terminated

One or more strands, or wires, left out of the terminal or socket will reduce the effective cross-sectional area of the conductor at that point. This may result in overheating because further resistance has been introduced into the circuit. The same effect could occur with a loose connection.

Terminating flexible cords or cables

Just the minimum amount of insulation should be removed to achieve an effective connection, with the terminal screw firmly clamping the conductor. A good, clean, tight termination is essential.

Insulation should preferably be removed with proprietary stripping tools as this will result in a cleaner, neater job and will avoid damaging the conductor, its insulation and sheath.

Method

1. Using side-cutting pliers or a knife, slit the outer sheath from the end.

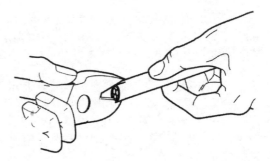

2. Peel back the two halves of insulated sheath for a suitable distance.

3. Cut off the sheath, neatly.
4. Examine the conductor insulation for damage.

Cable-stripping tools

1. Adjust the screw to suit the diameter of the conductor. Push the end of the wire into the tool so that the Vee slots close and cut through the insulation.

2. Remove the severed insulation.

3. Examine the conductor for damage.

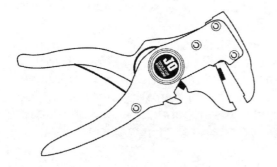 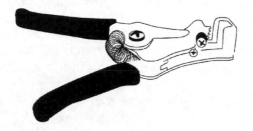

Cable-stripping tools

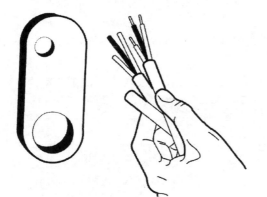

Flexible cable/wire stripper
and cutter

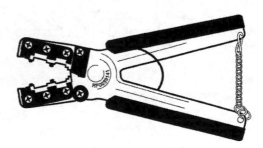

Twin and earth

Types of terminal

There is a wide variety of conductor terminations. The typical methods of securing conductors in accessories are pillar terminals, screwheads, and nuts and washers. Push-in connectors are also increasingly common.

Pillar terminals

A pillar terminal has a hole through its side into which the conductor is inserted and then secured by a set screw. If the conductor is small in relation to the hole, it should be doubled back. Care should be taken not to damage the conductor by excessive tightening.

Screwhead and nut and washer terminals

When fastening conductors under screwheads or nuts, it is best to form the conductor end into an eye, using round-nosed pliers. The eye should be slightly larger than the screw shank, but smaller than the outside diameter of the screwhead, nut or washers. The eye should be placed in such a way that rotation of the screwhead or nut tends to close the joint in the eye. If the eye is put the opposite way round, the motion of the screw or nut will tend to untwist the eye, and will probably result in imperfect contact.

Strip connectors

The conductors to be terminated are clamped by means of grub screws in connectors which are usually made of brass and mounted in a moulded, insulated or porcelain block.

Just the minimum amount of insulation should be removed to achieve an effective connection so that the terminal screw firmly clamps the conductor. A good, clean, tight termination is essential in order to avoid a high-resistance connection that could result in overheating of the joint.

Where possible, single conductors should always be folded.

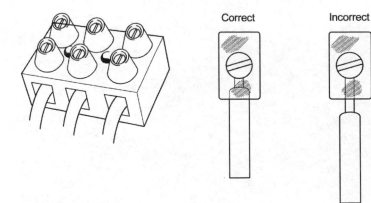

Correct   Incorrect

Terminating a plug to flexible cord

Method

1. Remove the plug-top.

2. Bring the flexible cord alongside the plug to measure the required amount of sheathing to be removed (or use the manufacturer's stripping guide).

3. Cut the conductors to the correct length.

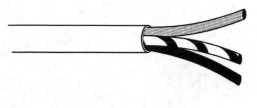

4. Remove sufficient insulation to expose the correct length of conductor to ensure a suitable termination.

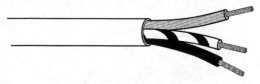

5. Make 'L' (brown), 'N' (blue) and 'E' (green/yellow) connections, ensuring that the conductors go to the correct terminals.

   Just the minimum amount of insulation should be removed to achieve an effective connection, and every strand or wire should be securely connected. Ensure that the terminal screws are tight.

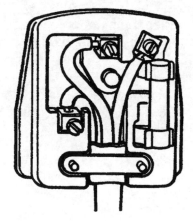

6. Clamp the flexible cord securely in or with the cord grip.

7. Replace the plug-top.

Terminating cable ends to crimp terminals

In order to terminate conductors effectively, crimp terminals are extensively used. This type of connection is often used in the termination of bonding conductors to earth clamps.

The terminals are usually made of tinned sheet copper with silver-brazed seams. The colour-coded crimp terminals represent the cable sizes they are designed for use with, and they are typically:

- RED $= 0.75$ mm² $- 1.5$ mm²
- BLUE $= 1.5$ mm² $- 2.5$ mm²
- YELLOW $= 4$ mm² $- 6$ mm²

Crimp terminals

The heavy-duty crimping tool is made with special steel jaws which are adjustable in order that a range of cables and terminals can be crimped.

The ratchet crimping tool must be fully closed before the jaws will open to release the crimp terminal. This is to ensure correct connection.

Heavy-duty crimping tool

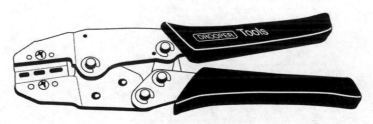

Ratchet crimping tool

Method

1. Remove the correct amount of cable insulation.

2. Place into the terminal.

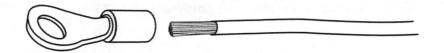

3. Crimp using a crimping tool in accordance with the manufacturer's instructions.

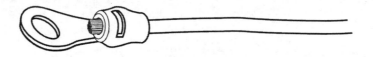

4. Check the connection for soundness by holding the cable firmly and giving the terminal a sharp tug between the thumb and forefinger.

PVC cable

PVC (polyvinyl chloride) insulated and sheathed cables are used extensively for lighting and heating installations in domestic dwellings, as they are generally the most economical method of wiring for this class of work.

Grades of PVC and their use in cords and cables and flexible cords

The PVC compounds used for cables and flexible cords are described in BS 6746 (1969). Several grades of compound are detailed in this standard for both insulation and sheathing requirements. PVC compounds are thermoplastic by nature and consequently are designed to operate within a prescribed temperature range. Grades of PVC can therefore be selected to suit a particular environmental temperature, the maximum conductor temperature for heat-resisting grade PVC being 85°C and the lowest operating temperature grade PVC being below minus 30°C.

The majority of wiring installations, however, use a general-purpose grade of PVC which has a maximum operating temperature of 70°C. This grade of PVC should not be installed when the air temperature is nearing 0°C.

Types of PVC cable and flexible cords

Single-core PVC insulated unsheathed cable

Application

Designed for drawing into conduit

Construction

PVC insulated solid or stranded copper conductor, coloured brown, black, grey, blue and green and yellow.

Single-core PVC insulated and sheathed cable with CPC

Application

For domestic and general wiring where a circuit protective conductor is required for all circuits.

Construction

PVC insulated plain copper conductor laid parallel with an uninsulated plain copper circuit protective conductor, sheathed overall with PVC compound. Core colour: brown. Sheath colour: grey.

PVC insulated and sheathed flat wiring cables

Application

For domestic and industrial wiring. Suitable for surface wiring where there is little risk of mechanical damage.

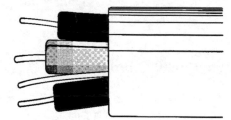

Construction

Two or three core cables. Two or three plain copper, solid or stranded conductors insulated with PVC and sheathed overall with PVC.

Core colours: two core: brown and blue; three core: brown, black and grey. Sheath colours: grey and white.

Note: White is now used for cable with low smoke (LSF) properties.

Single-core PVC insulated and sheathed cable

Application

Suitable for surface wiring where there is little risk of mechanical damage. Single-core is used for conduit and trucking runs where conditions are onerous.

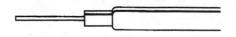

Construction

PVC insulated and PVC sheathed solid or stranded plain copper conductor. Core colours: brown or blue. Sheath colour: grey – other colours available.

PVC insulated and sheathed flexible cords

Application

General purpose indoors or outdoors in dry or wet locations. Portable tools, washing machines, vacuum cleaners, lawnmowers. Should not be used where the sheath can come into contact with hot surfaces. Not suitable for temperatures below 0°C.

Multicore versions of this cable up to 20 cores have uses in control equipment.

Construction

Two and three core cables exactly as before but with the inclusion of an uninsulated plain copper circuit protective conductor between the cores of twin cables and between the yellow and blue cores of three core cables.

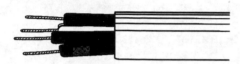

Heat resisting PVC insulated and sheathed flexible cords

Application

Suitable for use in ambient temperatures up to 45°C. Not suitable for use with heating appliances.

Construction

Plain copper flexible conductors, insulated with heat resisting (HR) PVC and HR PVC sheathed.

Core colours: single core: brown or blue; twin core: brown and blue; three core: brown, blue and green/yellow; four core: brown, black, grey and green/yellow. Sheath colour: white.

PVC insulated and sheathed flat twin flexible cord

Application

Intended for light duty indoors, for table lamps, radios and TV sets where the cable may lie on the floor; should not be used with heating appliances.

Construction

Plain copper flexible conductors PVC insulated, two cores laid parallel and sheathed overall with PVC.

Core colours: brown and blue. Sheath colour: white.

PVC insulated and sheathed light duty flexible cord

Application

Sometimes known as pendant flexibles, these are used for lighting fittings, push switches and other light domestic applications.

Construction

Two or three core cords having plain copper flexible conductors PVC insulated and sheathed.

Core colours: twin, brown and blue; three core: brown, blue and green/yellow. Sheath colour: white.

Flexible cords

A flexible cord is up to 4 mm^2 in size, a flexible cable is 4 mm^2 and above.

There are various types of flexible cord. The type required for a particular use depends on the type of equipment it is to be connected to and where it is to be used.

A guide to the correct selection of a flexible cord for a particular use is given in the Tables of the IEE On-Site Guide to the IEE Regulations.

The size of the flexible cord to be selected is determined by the current rating of the appliance. This is given in Tables in the IEE Wiring Regulations and is based on an ambient temperature of 30°C. Where the temperature is greater than this, correction factors must be applied.

10. INSTALLING PVC CABLES

Methods

The IEE regulations' requirements for installing cables is that if it is not continuously supported, as is the case when pulled into conduit or trunking installations, then the cables should be supported by suitable means at appropriate intervals to prevent mechanical strain in the terminations of the conductors and the conductors themselves. For PVC cables this requirement can be met by the use of plastic clips that incorporate a masonry nail.

Guidance on the spacing of clips for PVC insulated and sheath cables is specified in Table 4A of the IEE *On-site Guide* to the sixteenth edition of the IEE Wiring Regulations (an extract is given below):

	Support spacings	
Cable diameters (mm)	Horizontal (mm)	Vertical (mm)
Up to 9	250	400
9 to 15	300	400
15 to 20	350	450
20 to 40	400	550

In locations such as under the floor and behind partitions where cables are unlikely to be disturbed, greater distances can be used. It will usually be found necessary to fix clips closer together, especially on larger cables, if a neat appearance is to be achieved.

Surface wiring

Where PVC cables are on the surface, the cable should be run directly into the electrical accessory, ensuring that the outer sheathing of the cable is taken inside the accessory.

Concealed wiring

If the cable is concealed, a flush box is usually provided at each control or outlet position.

Clipping PVC cable

Ensure the electricity supply is isolated:

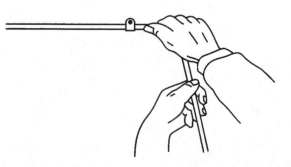

1. In order to ensure a neat appearance, PVC cable should be pressed flat against the surface between cable clips.

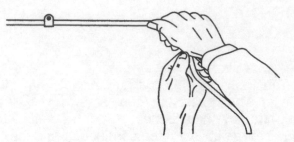

2. The cable should be formed by running the thumb against the surface of the cable, as illustrated.

3. Another method of forming the cable is to run the palm of the hand along the surface of the cable, as illustrated.

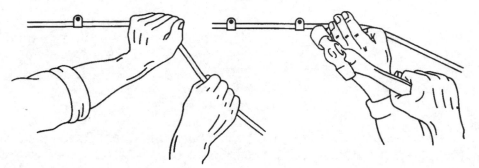

4. This sequence of forming the cable should be carried out after inserting the last cable clip and before fixing the next cable clip.

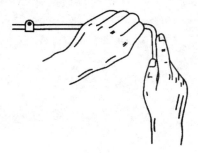

5. When a PVC cable is to be taken round a corner or changes direction, the bend should be formed using the thumb and fingers, as shown.

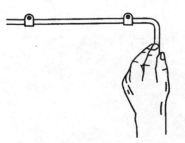

6. Care must be taken to ensure that the bend does not damage the cable or conductors. The cable must be supported at appropriate intervals so that it can support its own weight without damage.

Routing of PVC cables

Cables installed under the floors

Cables installed under floors and over ceilings must be routed so that they will not be damaged through contact with the floor or ceiling, or by the method of fixing. This involves the careful routing and clipping of cables.

Cables that are run in the space under floors and over ceilings should be installed at least 50 mm below the surface to prevent penetration by the nails or screws used in fixing flooring and ceiling materials. Alternatively, cable should be installed in an earthed steel conduit that is securely supported, or provided with equivalent mechanical protection that will prevent penetration by nails or screws, etc.

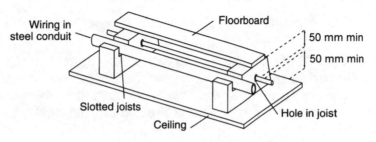

Support and protection for cables run under floorboards

Cables concealed in walls or partitions

When cables are installed in walls or partitions at depths of less than 50 mm, the risk of damage must be minimised. The permitted methods are as follows:

- the use of cables that have an earthed metallic covering that is suitable as a protective conductor (steel-wire armoured cable or MICC to BS 5467, 6346, 6724, 7846, 8436 or BS EN 60702-1)

- the use of cables that are installed in earthed conduit or trunking that is suitable as a protective conductor, or provided with adequate mechanical protection to prevent damage to the cable by screws, nails, etc.

- the cables are installed in a 150 mm zone from the top of a wall or partition, or within 150 mm of an angle created by adjoining walls or partitions

- the cables are run horizontally or vertically to accessories installed on walls or partitions. *Note: If the location of an accessory can be determined from the reverse side and the wall or partition is of ≤ 100 mm thickness, the zone extends to the reverse side of the wall or partition.*

103

In view of the practical problems, however, it is likely that cables will be installed mainly in the permitted zones.

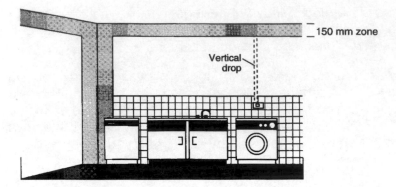

Cables should be run in permitted zones, or horizontally or vertically direct to the accessory

Even heavy-gauge steel conduit does not give complete protection against mechanical damage and so the cable may need to be replaced. A capping of metal or plastic is used to protect cables laid under plaster. This will protect the cables during the plastering operation, but gives very limited protection against nails and other objects driven into the plaster.

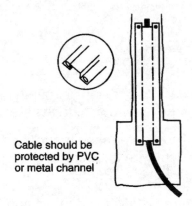

Cable should be
protected by PVC
or metal channel

Cables installed in floors, walls and ceilings

Wiring systems should only be concealed in floor or ceiling voids, or in internal wall spaces – not in external cavities – for the following reasons:

- PVC insulated cable installed in an external wall cavity may be adversely affected by the introduction of cavity wall insulation and this may result in the cable overheating

- the cable may also 'bridge' the air gap and cause moisture to be transmitted from the exterior to under the wall's surface, via the cable.

Other mechanical stresses (AJ)

Conductors and cables should be installed so that they are protected against any risk of mechanical damage.

Where cables pass through holes in metalwork, such as metal accessory boxes and luminaries, bushes or grummets must be fitted to prevent abrasion of the cables on any sharp edges.

Conductors and cables should not be subject to damage from incorrect bend radius, inadequate support (including damage from their own weight) or be subject to any excessive mechanical strain. Flexible wiring should not suffer from undue torsional or tensile stresses on terminations or conductors.

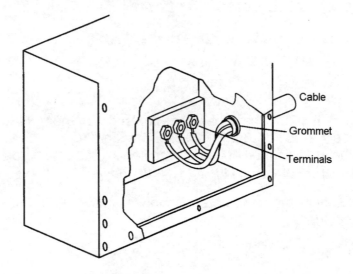

Reducing the spread of fire risk

Steps must be taken when installing cables to reduce the risk of spread of fire.

Where a wiring system is required to pass through or penetrate material that forms part of the construction of a building (for example, cable, trunking or busbar trunking systems), areas external to the wiring system and, where necessary, internal areas, must be sealed to maintain the required fire resistance of the material.

Wiring systems with non-flame-propagating properties and that have an internal cross-section not exceeding 710 mm need not be sealed internally. The sealing system used must meet the following requirements. It must:

- be compatible with the wiring system concerned

- permit the thermal movement of the wiring system without detriment to the sealing

- be removable without damage when additions to the wiring system are necessary

- be capable of resisting external influences to the same standards as the wiring system.

During the installation of wiring systems, temporary sealing arrangements must be made. In addition, any existing sealing disturbed or removed in the course of alterations to an installation must be reinstated as soon as possible.

Keeping cables away from other service installations

Care must be taken when installing PVC cables to ensure that they are not allowed to come into contact with gas pipes, water pipes and any non-earthed metalwork.

BS 7671 IEE REGULATIONS (16TH EDITION)

Harmonised cable core colour (Appendix 7)

This information also appears in the IEE *On-site Guide* (Appendix 11: Identification of Conductors).

Introduction

The requirements of BS 7671 have been harmonised with the requirements of the European Electrical Standards Body, CENELEC, regarding the identification of conductors and the identification of cores in cables and flexible cords. (See BS 7671, Tables 51 and 7A–7E, or the IEE *On-site Guide*, Tables 11A–11F.)

These standards specify the cable core marking, including the cable core colours, for the CENELEC countries.

For single-phase installations:

The RED phase and BLACK neutral

are replaced by

BROWN phase and BLUE neutral

The protective conductor remains GREEN and YELLOW

Therefore, fixed wiring for single-phase installations will now adopt the same colours that three-core flexible cables and cords have used for many years.

Notes: Installations beginning after 31 March 2004 may use the existing or the new harmonised core colours, but not both. Installations starting after 31 March 2006 must use only the new harmonised colours.

Wherever an interface (connection) occurs between old and new cable colours, a warning notice should be displayed.

CAUTION
This installation has wiring colours to two versions of BS 7671.
Great care should be taken before undertaking extension,
alteration or repair that all conductors are correctly identified.

Switch wires

New installations or modifications to existing installations

Where a PVC-insulated and sheathed (twin and earth) cable is used as a switch wire, and both cores are used as phase conductors, these are coloured BROWN and BLUE:

- the BLUE conductor must be oversleeved BROWN or marked L at its terminations

- the bare CPC must be oversleeved GREEN and YELLOW as normal.

Use of three-core and CPC cable

New installations or modifications to existing installations

Where a three-core and earth cable with core colours BROWN, BLACK and GREY is used to wire a room thermostat:

- the GREY conductor should be oversleeved BROWN or marked L at its terminations

- the bare CPC must be oversleeved GREEN and YELLOW as normal.

Phase conductors

New installations or modifications to existing installations

The colour and markings of phase conductors should be as BS 7671, Table 51.

For control circuits, extra-low voltage and other applications:

- the phase conductor could be coloured BROWN, BLACK, RED, ORANGE, YELLOW, VIOLET, GREY, WHITE, PINK or TURQUOISE and marked L

- the neutral should be coloured BLUE and marked N.

An earthed, protective, extra-low voltage (PELV) conductor must be BLUE.

Protective conductors

- Any bare CPCs incorporated in cables, e.g. twin and earth, must be exclusively used as a protective conductor, oversleeved green and yellow and **MUST NOT** be used for any other function or purpose

- The two colour combination s green and yellow must be used exclusively for the identification of protective conductors and this colour combination **MUST NOT** be used for any other purpose

- Single core cables or the cores of flexible cords that are coloured green and yellow must be exclusively used as protective conductors and **MUST NOT** be oversleeved at their terminations and used for other functions, e.g. switch wire from a room thermostat or cylinder thermostat that does not require earthing.

Alterations or additions to existing installations

Single-phase installations

Alterations or additions to a single-phase installation do not need marking at the interface where old wiring is connected to new, providing that:

- the old cables are coloured RED for phase and BLACK for neutral, AND

- the new cables are coloured BROWN for phase and BLUE for neutral.

Two-phase or three-phase installations

At the wiring interface between old core colours and new core colours, clear identification is required, as follows:

Old and new conductors

Neutral conductors = **N**

Phase conductors = **L1, L2, L3**

BS 7671, Table 7A, gives examples of conductor markings at an interface for additions and alterations to an AC installation that is identified with the old cable colours (see below).

Function	Old conductor colour	Old/new marking	New conductor colour
Phase 1	RED	L1	BROWN
Phase 2	YELLOW	L2	BLACK
Phase 3	BLUE	L3	GREY
Neutral	BLACK	N	BLUE
Protective conductor	GREEN & YELLOW		GREEN & YELLOW

For a three-phase installation (as an alternative to the BROWN, BLACK and GREY identification of the phase conductors shown above), three BROWN or three BLACK or three GREY conductors may be used. However, they must be marked L1, L2 and L3 or oversleeved BROWN, BLACK and GREY at their terminations.

11. ELECTRICAL SAFETY CHECKS

Introduction

When it is properly used and maintained, electrical equipment has a high degree of safety but, as with any other technical equipment, misuse and neglect can lead to unnecessary safety hazards. For this reason, it is essential to check electrical equipment before it is used. The extent to which you must check equipment will depend on the type of equipment you are about to use and the nature of the work you are about to carry out. This chapter provides guidance on the checks and tests that may be necessary when you are working with electrical equipment.

Portable tools and appliances connected via flexible cords are, by their very nature, subject to mechanical damage, unauthorised repairs and variable operating conditions. At the very least, you should always carry out the following user checks whenever you start to use a piece of equipment or move the equipment to a new situation. You may also need to follow additional electrical safety precautions if the situation or your company's procedures demand them or if you feel they would provide a higher degree of safety.

User checks

Each time you use the equipment, ask yourself the following questions:

Is it suitable for the PURPOSE? Electrical equipment is no different from any other tool in this respect. Ask yourself the question – is this the right tool for the job? Do not be tempted to use improvised tools that may not be up to the job.

Is it suitable for the LOCATION? Where are you using the equipment? Even simple battery-operated equipment can present a hazard if used in the wrong environment – a spark from a torch could cause a gas explosion, for example.

Is it UNSAFE? A simple visual check should be carried out to verify that the equipment has not been damaged since the last time it was used. If it has been damaged, do not be tempted to carry out a temporary repair; get it fixed properly or withdraw it from service. Even new equipment cannot be relied upon to remain in a safe condition – it may have been damaged in transit, for example. It is estimated that up to 95% of electrical faults can be found by a visual inspection.

Is there a GREEN STICKER? This might seem strange, but many companies operate a system of regular portable appliance testing (PAT). Appliances that have passed the PAT test are often identified by a green (or yellow) sticker indicating that, at some time, they passed an electrical safety test. Look for the green sticker.

Is the SUPPLY OK? Is the supply you are connecting to of the right type (i.e. is it the right voltage, current and frequency)? Is the socket undamaged? Is it overloaded?

To help you remember these checks, notice that the first letters of the words in capital letters above spell the mnemonic **PLUGS** – **P**URPOSE, **L**OCATION, **U**NSAFE, **G**REEN STICKER, **S**UPPLY. So before you plug in, use PLUGS!

Electrical safety precautions

If you use electrical equipment in a variety of situations or at various customers' sites, or if you are involved in installing equipment that needs an electricity supply, there are two useful precautions you can take in addition to the user checks outlined above.

Supply check

On a strange site you may feel that the supply has not been wired correctly and therefore that it is not in a safe condition. The use of a socket outlet tester can provide a very quick and easy check of the supply simply by plugging it in. This is no substitute for a full electrical test, but it does at least show that all three wires are connected and that they are in the right places. If the supply fails this test, you or anyone else must not use it until a competent person has corrected the fault. It is a good idea to label the outlet as being unsuitable for use before reporting it to the appropriate person.

Note: These devices are not capable of identifying neutral-earth reversals.

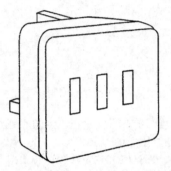

'Martindale' socket outlet tester

RCD protection

When you use portable equipment, a 30 mA residual current device (RCD) can provide additional protection against electric shock, and an RCD must be used to protect equipment that is being employed outdoors. The portable plug-in type (illustrated below) provides the most flexibility for on-site use. The device must be regularly tested to ensure that its effectiveness is maintained. Pressing the RCD test button is an easy way to check that the mechanism is operational, and the device should also be tested electrically on regular occasions to ensure that its electrical performance is within limits.

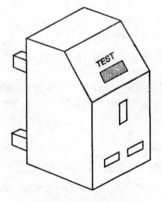

'Plug-in' type RCD

Electrical safety tests

If you are involved in connecting and reconnecting electrical equipment you will need to carry out more formal electrical safety tests as part of the commissioning process. These tests must be performed in the order given and, unless the test requires it, with the equipment isolated from the electrical supply.

Inspection

The following items (where applicable) should be inspected to make sure that the work has been carried out correctly, using the appropriate materials:

- connection of conductors – are all connections mechanically sound and in the right terminals?

- identification of conductors – are the cores the correct colours and suitably labelled?

- routing of conductors – are these subject to mechanical damage?

- selection of conductors – are these the right size?

- insulation – are all live parts suitably covered?

- enclosures – are all live parts inaccessible?

- protective conductors – is everything earthed that needs to be?

- bonding conductors – is everything bonded that needs to be?

- external influences – is there protection against water, heat, smoke, fumes, dust, etc.?

Please note that this is intended to be a guide only. There may be other factors that need to be taken into account because each situation will be different.

Earth continuity check

This check is carried out after safe isolation. It ensures that all the metallic parts of the system are satisfactorily earthed and that all earth connections are good.

Set the multimeter on the lowest ohms range. Connect one lead of the test meter to the earth pin on the plug or, if not connected by a plug, then the main incoming earth terminal. The other test meter lead should now be moved around the system to make a connection with all the earth terminals and the metallic components involved (for example, boiler casing, pipework, pump, etc.). In all instances the resulting reading should be less than one ohm. A reading in excess of one ohm indicates a poor earth connection, and this must be investigated.

Short circuit check

This check is also carried out after safe isolation. It ensures that there are no short circuits between live and neutral.

Set the multimeter on the lowest ohms range and ensure that all switches and controls are calling for heat. Connect one meter lead to the main incoming neutral terminal. The other lead should now be connected, in turn, to all the live connections on the terminal strip (this ensures that all the components are tested even if they are not switched in). A reading of less than one ohm indicates a short circuit.

Resistance to earth check

This check is again carried out after safe isolation. It tests the quality of the insulation of the wiring and components in the system. A failure on this check means the system is dangerous and therefore it must not be connected to the electrical supply until the problem has been rectified.

Set the multimeter on the highest ohms range and ensure that all switches and controls are calling for heat. Connect one meter lead to the main incoming earth terminal. The other lead should now be connected, in turn, to all the live connections on the terminal strip (this ensures that all the components are tested even if they are not switched in). An infinitely high reading should be achieved in all instances. Any reading less than 2 MΩ must be investigated.

Mains voltage and polarity check

This test is carried out with the power to the system switched on. All the safety precautions called for when testing live must be observed. This test checks the main incoming voltage to the system and ensures that the live, neutral and earth terminals are connected.

Set the multimeter to an AC voltage range suitable to read 230 V, ensure that the instrument test leads and probes comply with GS 38 then carry out the following three checks:

1. read between live and neutral – the result should be 230 V AC

2. read between live and earth – the result should be 230 V AC

3. read between neutral and earth – the result should be between 0 V AC and 15 V AC.

If the voltage 0 V AC to 15 V AC occurs on any terminals other than neutral and earth, there is an electrical fault (i.e. incorrect polarity).

CITB constructionskills

BES PUBLICATIONS

Building Engineering Services continue to provide the gas, electric, water and refrigerant industries with a range of popular, respected and competitively priced publications.

These publications can be used either as the basis of training or for reference in the workplace. Some can also be used for assessment purposes. All are published in A4 format, with the most popular also available as A5, pocket-sized books.

DOMESTIC GAS

GAS SAFETY (G1)   Price £50.00   Format: A4 in a ringbinder
Over 300 pages of clear explanation and illustrations covering the essentials in ◆gas pipework ◆gas supply ◆combustion ◆appliance gas safety devices and gas controls ◆principles of gas flues ◆flueing standards ◆ventilation requirements ◆emergency procedures ◆unsafe situations ◆warning notices and labels.
Also included is the HSE publication ◆*Safety in the installation and use of gas systems and appliances* (G31) which covers the HSE Gas Safety (Installation and Use) Regulations 1998 – Approved Code of Practice and Guidance, a ◆*Course Workbook* and a booklet of ◆*Practical Tasks* for you to complete.

GAS SAFETY (G2)   Price £28.00   Format: A5 Wiro-bound
All the information and diagrams from the *GAS SAFETY (G1)* package in a handy size for reference on the job and for carrying in the service van.

DOMESTIC GAS APPLIANCES (G5)   Price £70.00   Format: A4 in a ringbinder
All the CITB-ConstructionSkills Domestic Natural Gas Appliance manuals in one plus the *Domestic Natural Gas Appliance Course Workbook*. The easy-to-use format makes it ideal for those working over a range of domestic appliances.
Each manual can also be purchased individually:
* Domestic Heating Boilers/Water Heaters (G7) £14.00
* Domestic Cookers (G8) £12.00
* Domestic Ducted Air Heaters (G9) £12.00
* Domestic Fires and Wall Heaters (G10) £14.00
* Domestic Tumble Dryers (G11) £10.00
* Domestic Meters (G12) £10.00
* Domestic Instantaneous Water Heaters (G13) £14.00

DOMESTIC GAS APPLIANCES (G6)   Price £38.00   Format: A5 Wiro-bound
All the information and diagrams from the *DOMESTIC GAS APPLIANCES (G5)* package in a handy size for reference on the job and for carrying in the service van.

SAFETY AT WORK (G15)   Price £15.00   Format: A4
Essential advice on safety at work, from securing ladders to dealing with electric shock. It also gives the key points of relevant Acts and regulations.

FAULT-FINDING TECHNIQUES (G17)   Price £15.00   Format: A4
Problems with locating that elusive fault? Follow the step-by-step techniques in this hands-on manual and speed up your fault finding on central heating systems.

SAFETY IN THE INSTALLATION AND USE OF GAS SYSTEMS AND APPLIANCES (G31) Price £10.95   Format: A4
An essential HSE publication for all those working with domestic gas. It gives advice on how to comply with *The Gas Safety (Installation and Use) Regulations 1998 – Approved Code of Practice and Guidance* which has a special legal status. For example, if you are prosecuted for breach of health and safety law, and it is proved that you have not followed the relevant provisions of the Code, a court will find you at fault (unless you can show that you have complied with the law in some other way).

LIQUEFIED PETROLEUM GAS (LPG)

LIQUEFIED PETROLEUM GAS SAFETY (G80)   Price £50.00   Format: A4 in a ringbinder
The industry reference manual for those working only on LPG systems. It covers all you need to know about ◆combustion ◆appliance gas safety devices and gas controls ◆flueing standards and materials ◆flue systems ◆ventilation requirements ◆emergency procedures ◆unsafe situations ◆warning notices and labels.
Also included is the HSE publication ◆*Safety in the installation and use of gas systems and appliances* (G31) which covers the HSE Gas Safety (Installation and Use) Regulations 1998 – Approved Code of Practice and Guidance, and a ◆*Course Workbook*.

LIQUEFIED PETROLEUM GAS (G18)   Price £15.00   Format: A4
The essential bolt-on to those working with natural gas and looking to extend into LPG. If you already own a *GAS SAFETY (G1)* pack, all you need is this book with its LPG-specific sections: ◆installation ◆fire precautions and procedures ◆combustion ◆testing and commissioning installations ◆service pipework ◆bulk gas supply systems ◆the leisure industry.

Prices are correct at the time of going to press. To obtain further information and order any of the publications listed, contact BES Publications Sales on: Tel: 01485 577704 / Fax: 01485 577713 / E-mail: bes.enquiry@citb.co.uk

COMMERCIAL AND INDUSTRIAL GAS

COMMERCIAL GAS SAFETY (G88) Price £75.00 Format: A4 in a ringbinder
An essential training and reference manual for those working in the commercial environment. It builds on the popular GAS SAFETY (G1) and incorporates information from two other commercial publications (sold separately) making this the definitive training and reference manual for commercial work, it covers ◆commercial gas safety ◆pipework and ancillary equipment ◆gas supply ◆combustion ◆burners ◆controls ◆flues ◆ventilation ◆pressure and flow ◆emergency procedures ◆combustion and flue gas analysis ◆pipework design ◆soundness testing and purging ◆commercial metering ◆boosters and compressors. Also included is the HSE publication ◆*Safety in the installation and use of gas systems and appliances* (G31) a ◆*Course Workbook* and a booklet of ◆*Practical Tasks.*

COMMERCIAL GAS SAFETY (G23) Price £20.00 Format: A4
An essential supplement for engineers working in the commercial environment. If you already own a *GAS SAFETY (G1)* pack, all you need is this book with its commercial gas specific sections: ◆combustion and flue gas analysis ◆burners ◆controls and control systems ◆flues ◆ventilation ◆pressure and flow.

COMMERCIAL PIPEWORK AND ANCILLARY EQUIPMENT (G24) Price £20.00 Format: A4
An essential guide for engineers working on commercial pipework. There is clear information on ◆pipework design ◆soundness testing and purging ◆commercial metering ◆boosters and compressors.

COMMERCIAL APPLIANCES (G25) Price £15.00 Format: A4
A comprehensive guide to the installation and commissioning of direct and indirect fired appliances, radiant heating and gas equipment.

COMMERCIAL CATERING (G26) Price £15.00 Format: A4
Essential information on the installation, commissioning and servicing of commercial catering appliances.

ELECTRICAL

IEE REGULATIONS – 16TH EDITION (E1) Price £16.00 Format: A4 Wiro-bound
The standard reference book for all engaged in electrical work. The easy-to-follow text, supported by diagrams, explains the complex regulations in terms a practical electrician can understand. And it now incorporates reference to the IEE *On-site Guide* that enables you to make calculations and design circuits in a much quicker and simpler manner.

ELECTRICAL INSTALLATION PACK (E3) Price £48.00 Format: A4 in a ringbinder
Over 430 pages of illustrated reference material divided into four sections:
- Basic Practical Skills – describes the tools required for electrical installation work and how to use them
- Wiring Installation Practice – deals with terminating cables, flexible cords and installing PVC cables, conduit trunking, MICC, SWA and FP200 wiring systems (complies with the 16th edition IEE Wiring Regulations)
- Basic Electrical Circuits – covers standard circuit arrangements for lighting and power circuits, and relevant IEE regulations
- Safety at Work – essential advice on safety at work, from securing ladders to dealing with electric shock. It also gives the key points of relevant Acts and regulations.

ESSENTIAL ELECTRICS (E14) Price £18.00 Format: A4
An essential reference book for plumbers, gas fitters and heating and ventilating engineers whose work requires basic electrical knowledge and an understanding of electrical regulations.

CENTRAL HEATING CONTROLS (E15) Price £12.00 Format: A4
Covers the different types of central heating control systems involving the components: ◆ wiring ◆ fault finding.

COMBINATION BOILERS (E19) Price £20.00 Format: A4
A valuable reference manual for engineers who want to understand the principles of combination boilers. This manual also forms the backbone of CITB-ConstructionSkills 'Essential Electrics Combination Boiler fault finding course'. Over 80 pages of illustrated reference information covering: ◆ types of boilers ◆ designs ◆ wiring diagrams ◆ installation ◆ commissioning and servicing ◆ fault diagnosis.

WATER

UNVENTED HOT WATER STORAGE SYSTEMS (UHWSS) (W2) Price £15.00 Format: A4
A comprehensive guide covering ◆types of system ◆design ◆controls ◆installation ◆commissioning ◆servicing ◆fault diagnosis ◆decommissioning ◆The Building Regulations ◆good practice.

REFRIGERANTS

SAFE HANDLING OF REFRIGERANTS (R2) Price £15.00 Format: A4
Essential information for anyone handling refrigerants, primarily designed for operatives undertaking Safe Handling of Refrigerants assessments. Covers ◆environmental impact ◆fluorocarbon control and alternatives ◆regulations ◆good practice.

SAFE HANDLING OF ANHYDROUS AMMONIA (R4) Price £18.00 Format: A4
Essential information for anyone handling anhydrous ammonia, primarily designed for operatives undertaking Safe Handling of Anhydrous Ammonia assessments. Covers ◆safety and environmental issues ◆regulations ◆good practice.

PIPEWORK AND BRAZING (R6) Price £16.00 Format: A4
Primarily for operatives undertaking CITB-ConstructionSkills Pipework and Brazing assessments for refrigeration systems. Covers ◆health and safety ◆materials and equipment ◆lighting procedures.